A. Avantaggiati (Ed.)

Pseudodifferential Operators with Applications

Lectures given at a Summer School of the
Centro Internazionale Matematico Estivo (C.I.M.E.),
held in Bressanone (Bolzano), Italy,
June 16-24, 1977

Springer

FONDAZIONE
CIME
ROBERTO CONTI

C.I.M.E. Foundation
c/o Dipartimento di Matematica "U. Dini"
Viale margagni n. 67/a
50134 Firenze
Italy
cime@math.unifi.it

ISBN 978-3-642-11091-7 e-ISBN: 978-3-642-11092-4
DOI:10.1007/978-3-642-11092-4
Springer Heidelberg Dordrecht London New York

Reprint of the 1st ed. C.I.M.E., Ed. Liguori, Napoli 1978
With kind permission of C.I.M.E.

Printed on acid-free paper

Springer.com

CENTRO INTERNAZIONALE MATEMATICO ESTIVO

(C.I.M.E.)

II Ciclo - Bressanone dal 16 al 24 giugno 1977

PSEUDODIFFERENTIAL OPERATORS WITH APPLICATIONS

Coordinatore: Prof. A. Avantaggiati

CENTRO INTERNAZIONALE MATEMATICO ESTIVO

(C.I.M.E.)

PSEUDO-DIFFERENTIAL OPERATORS ON HEISENBERG GROUPS

A. DYNIN

Corso tenuto a Bressanone dal 16 al 24 giugno 1977

Introduction

The convolution operators on the euclidean spaces are only a particular case of convolution operators on arbitrary Lie groups. Pseudo-differential operators are roughly speaking convolution operators with variable coefficients. In classical theory of such operators it is important to have standard dilations on a euclidean space. So we pay attention only to Lie groups with dilation i. e. with 1-dimensional group of automorphisms converging to infinity when real parameter increases to infinity.

It is well known that all Lie groups with dilations are nilpotent.

In this seminar I consider only Heisenberg Groups. These groups are the simplest non-abelian nilpotent groups, They appear in Complex Analysis and such operators on strictly pseudoconvex boundaries as J. Kohn sub-Laplacian, induced Cauchy-Riemann operators, singular integral operators of Cauchy-Henkintype can be locally considered as convolution operators wtih variable coefficients on Heisenberg groups. The characteristic feature of all these operators is an anisotropy of their singularities tight with complex tangent directions. Actually this is a contact structure. Generally a contact structure is given on 2n+1 -manifold by 1-form ω such that the (2n+1) -form $\omega \wedge d\omega \wedge \ldots \wedge d\omega \neq 0$

If a strictly pseudoconvex boundary is defined by real function ρ with

$d\rho \neq 0$ then we may take $\omega = \frac{1}{i} (\partial\rho - \overline{\partial\rho})$. By D-arboux theorem Heisenberg groups are local models for any contact manifolds.

I construct a theory of pseudo-differential operators which belong to the contact structure as classical pseudo-differential operators belong to the smooth structure.

It is impossible to derive a theory of pseudo-differential operators without an elliptic accompaniment. I introduce a kind of ellipticity which as I hope can elucidate some striking analogies between non-elliptic in usual sense $\overline{\partial}_b$-complex and elliptic complexes.

Note that the problem of pseudo-differential operators on homogeneous Lie groups was put forward by E. Stein at the Nice congress (cf /4/).

Our results were mainly announced in /2/ and /3/. Here we exposed them in more precise form.

1 Heisenberg Lie groups and algebras

Heisenberg algebra H_n, $n > 0$, can be obtain from the standard euclidean space $\mathbb{R}^{2n+1}$ if we supply it with such commutators:

if $x = (x_0, x^1, x^{11}) \in \mathbb{R}^{2n+1}$ where $x_0 \in \mathbb{R}$, $x^1, x^{11} \in \mathbb{R}^n$

$e_0 = (1; 0, 0,) \in \mathbb{R}^{2n+1}$ then

$$[x, y] = 4 (\langle x^1, y^{11}\rangle - \langle x^{11}, y^1\rangle) e_0$$

This Heisenberg algebra is a Lie algebra of the step 2 all $[x,[y,z]]$ are zero. Let $\mathbb{H}_n$ be a corresponding simply-connected Lie group. As manifold it is identified with $\mathbb{R}^{2n+1}$; the multiplication is

$$x * y = (x_0 + y_0 + 2(\langle x', y''\rangle - \langle x'', y'\rangle), x' + y', x'' + y'')$$

Therefore O serves as unit and $x^{-1} = -x$ The group $\mathbb{H}_n$ is called the Heisenberg group.

The identity mapping $x \mapsto x$ coincides with the exponential

$$\exp: H_n \to \mathbb{H}_n$$

There are dilations

$$\delta_t x = (t^2 x_o, tx^1, tx^{11}) \quad , \quad t > 0,$$

in the Heisenberg group $\mathbb{H}_n$. Of course $\delta_{t_1+t_2} = \delta_{t_1} \delta_{t_2}$ and the δ_t are automorphisms of the Lie structures in $\mathbb{H}_n$ and H_n. Let $s(\mathbb{H}_n)$, $s'(H_n)$ be corresponding spaces on $\mathbb{R}^{2n+1}$.

The operators of the left and of the right shifts by elements $y \in \mathbb{H}_n$

$$L(y) : x \mapsto y^{-1} \times x, \quad R(y) : x \mapsto x \times y, \quad x \in \mathbb{H}_n,$$

are continuous in this spaces. Moreover the Lebesgue measure is bilaterally invariant on the $\mathbb{H}_n$

We may identify the Heisenberg algebra H_n with the Lie algebra of left-invariant 1-st order differential operators. Pick out generators of the complexification of this algebra

$$X_o = \frac{1}{i} \frac{\partial}{\partial x_o}, \quad X_j' = \frac{1}{i} \left(\frac{\partial}{\partial x_j'} - 2x_j'' \frac{\partial}{\partial x_o} \right)$$

$$X_j'' = \frac{1}{i} \left(\frac{\partial}{\partial x_j''} + 2 x_j' \frac{\partial}{\partial x_o} \right), \quad j = 1, \dots, n$$

Adopt the notation

$$X' = (X_1', \dots, X_n'), \quad X'' = (X_1'', \dots X_n''), \quad X = (X_o, X', X'')$$

The contact structure on the $\mathbb{H}_n$ is defined by a left-invariant 1-form

$$\omega(x) = dx_o + 2 \langle x', dx'' \rangle - 2 \langle x'', dx' \rangle$$

3. H. Weyl quantization.

The modern theory of pseudo-differential operators took its shape in the sixties however we can find its origin as early as in the beginning of the thirties: there was a problem of quantization in the Quantum Mechanics and H. Weyl gave a general solution. The problem is to construct of non-cummuting operators of multiplication:

$$\hat{x} : u \mapsto x u \; ; \quad x = (x_1, x_2, \ldots, x_n)$$

and of differentiation

$$\hat{p} : u \mapsto \frac{1}{i} \frac{\partial}{\partial x} u , \quad \frac{\partial}{\partial x} = \left(\frac{\partial}{\partial x_1}, \ldots, \frac{\partial}{\partial x_n} \right).$$

This operators generate a 2-step Lie subalgebra W_n in the Lie algebra $\mathcal{L}(\mathcal{S}(\mathbb{R}^n))$ of all continuous linear operators in the $\mathcal{S}(\mathbb{R}^n)$.

Let $\mathbb{W}_n$ be the symple-connected Lie group of this W_n. Then the $\mathbb{W}_n$ can be realized as a Lie group of continuous linear operators in the $\mathcal{S}(\mathbb{R}^n)$. We have the exponential map $\exp : W_n \mapsto \mathbb{W}_n$

Therefore for a $f \in \mathcal{S}(\mathbb{R}^{2n}_{x,p})$ we can define an operator

$$(*) \quad f(\hat{x}, \hat{p}) = (2\pi)^{-2n} \int \tilde{f}(\xi, \eta) \exp i \left[\langle \xi, \hat{x} \rangle + \langle \eta, \hat{p} \rangle \right] d\xi \, d\eta$$

where

$$\tilde{f}(\xi, \eta) = \int e^{i[\langle x, \xi \rangle + \langle p, \eta \rangle]} f(x,p) \, dx \, dy$$

is the Fourier transform. The expression $(*)$ is similar to the inverse Fourier transform of the $\tilde{f}(\xi . \eta)$.

Let us justify this definition.

It is easy to see that the integral $(*)$ converges in the space $\mathcal{L}(\mathcal{S}(\mathbb{R}^n))$ under the strong topology and actually is an integral operator with the Schwartz kernel

$$\mathcal{K}_f(x,y) = \frac{1}{(2\pi)^{2n}} \int e^{i\langle \xi, \frac{x+y}{2}\rangle} \tilde{f}(\xi, y-x)\, d\xi .$$

We can rewrite this formula as follows

$$\mathcal{K}_f(x,y) = \mathcal{F}^{-1}_{p\to y-x} f\left(\frac{x+y}{2}, p\right)$$

where $\mathcal{F}^{-1}_{p\to z}$ is the inverse Fourier transformation (from the p to the z)

In this form the definition of $\mathcal{K}_f$ for every $f\in\mathcal{S}'(\mathbb{R}^{2n}_{x,p})$ is valid. If $f\in\mathcal{S}(\mathbb{R}^{2n}_{x,p})$ then $f(\hat{x},\hat{p})$ is a continuous operator from $\mathcal{S}'(\mathbb{R}^n)$ to $\mathcal{S}'(\mathbb{R}^n)$

If $f\in\mathcal{S}'(\mathbb{R}^{2n}_{x,p})$ then $f(\hat{x},\hat{p})$ is a continuous operators from $\mathcal{S}(\mathbb{R}^n)$ to $\mathcal{S}'(\mathbb{R}^n)$

Chose moderate operators (cf /5/): Let $m\in\mathbb{R}$. Take (cf. [5])

$W^m(\mathbb{R}^n) =$

$$\left\{ f\in C^\infty(\mathbb{R}^{2n}_{x,p}) : \forall\, \alpha,\beta\in\mathbb{Z}^m_+ ,\ \exists \text{ constant } C_{\alpha,\beta}\right.$$

such that $\left.\left|\left(\frac{\partial}{\partial x}\right)^\alpha\left(\frac{\partial}{\partial p}\right)^\beta f(x,p)\right| \le C_{\alpha,\beta}(1+|x|+|p|)^{m-|\alpha|-|\beta|}\right\}.$

We say that such f are symbols of **order** m

let

$$O_p W^m = \left\{ f(\hat{x},\hat{p}) : f\in W^m(\mathbb{R}^n)\right\}$$

If $f\in W^m(\mathbb{R}^n)$ then $f(x,\hat{p}) : \mathcal{S}(\mathbb{R}^n)\to\mathcal{S}(\mathbb{R}^n),\ \mathcal{S}'(\mathbb{R}^n)\to\mathcal{S}'(\mathbb{R}^n)$

and these operators are called Weyl operators. In particular it is possible to take a product of Weyl operators If $f_j\in O_p W^{m_j}$, $j=1,2$, then there exists a $f\in W^{m_1+m_2}(\mathbb{R}^n)$ such that

$$f(\hat{x}, \hat{p}) = f_1(\hat{x}, \hat{p})\, f_2(\hat{x}, \hat{p})$$

and

$f(x, p) = (f_1 f_2)(x, p) + \frac{1}{2}\{f_1, f_2\}(x, p)$ + terms of the lower order. This a simple consequence of the Hausdorff-Campbell formula for the product of exponents.

Moreover

$$f(\hat{x}, \hat{p})^* = \bar{f}(\hat{x}, \hat{p})$$

Let $W^m_o(\mathbb{R}^n)$ be the subspace of $W^m(\mathbb{R}^n)$ whose elements have the following property: there exists $f_o \in W^m(\mathbb{R}^n)$ such that

(i) $f_o(tx, tp) = t^m f_o(x, p)$, $\forall\, t > 1$, $|x| + |p| > 1$

(ii) $f - f_o \in W^{m-1}(\mathbb{R}^n)$

We say that such f are symbols with principal part f_o.

An operator $f(\hat{x}, \hat{p}) \in O_p W^m_o$ is called <u>elliptic</u> if $f_o(x, p) \neq 0$ for $|x| + |p| > 1$; it is known that the elliptic operators are hypoelliptic (cf /5/). Let $Ell\, W^m_o$ be the class of all elliptic operators of order m. If $f(\hat{x}, \hat{p}) \in Ell\, W^m_o$ then there exists its parametrix $r(\hat{x}, \hat{p}) \in Ell\, W^{-m}_o$ with principal part $r_o(x, p) = (f_o(x, p))^{-1}$ for $|x| + |p| > 1$

<u>Example</u>: the Hermite operator $E = -\frac{\partial^2}{\partial x^2} + x^2 \in Ell\, W^2_o$.

Let

$$\mathcal{E}^k(\mathbb{R}^n) = \{u \in \mathcal{S}'(\mathbb{R}^n) : \|u\|_k = \|E^{k/2} u\|_{\mathcal{J}^2} < \infty\}$$

Then $f(\hat{x}, \hat{p}) \in Ell\, W^m_o$ if and only if $f(\hat{x}, \hat{p})$ is a Fredholm operator from $\mathcal{E}^k(\mathbb{R}^n)$ to $\mathcal{E}^{k-m}(\mathbb{R}^n)$ for some (and therefore for any) $k \in \mathbb{R}$.

4. Pseudo - differential operators on $\mathbb{H}_n$

Apply the Weyl quantization for construction of pseudo-differential operators on $\mathbb{H}_n$.

First of all the operators of multiplication by coordinate functions.

$$\hat{x} : u(x) \to x\, u(x)\ , \qquad x = (x_0, x', x'')$$

and left- invariant operators

$$\hat{X} : u \to X u \qquad\qquad X = (X_0, X', X'')$$

generate a 3-step Lie subalgebra in Lie algebra $\mathcal{L}(\mathcal{S}(\mathbb{H}_n))$. Treating this Lie algebra as above enables to define

$f(\hat{x}, \hat{X}) =$

$$(**) \qquad \frac{1}{(2\pi)^{4n+2}} \int \tilde{f}(\xi, \eta) \exp i[\langle \xi, \hat{x} \rangle + \langle \eta, \hat{X} \rangle]\, d\xi\, d\eta$$

If $f \in \mathcal{S}(\mathbb{R}^{4n+2}_{x,X})$ then $(**)$ is an integral operator with the Schwartz Kernel

$$\mathcal{K}_f(x,y) = \mathcal{F}^{-1}_{X \to x * y^{-1}} f\left(\frac{x+y}{2}, X\right).$$

This formula is valid for every $f \in \mathcal{S}'(\mathbb{R}^{4n+2}_{x,X})$ and we can use it for definition of $f(\hat{x}, \hat{X})$

If $f \in \mathcal{S}(\mathbb{R}^{4n+2}_{x,X})$ then $f(\hat{x}, \hat{X}) : \mathcal{S}'(\mathbb{H}_n) \to \mathcal{S}(\mathbb{H}_n)$

If $f \in \mathcal{S}'(\mathbb{R}^{4n+2}_{x,X})$ then $f(\hat{x}, \hat{X}) : \mathcal{S}(\mathbb{H}_n) \to \mathcal{S}'(\mathbb{H}_n)$

Let $[y]$ be a γ_t - homogeneous function on the $\mathbb{H}_n$

$$[y] = (y_0^2 + (y'^2 + y''^2)^2)^{1/4}$$

For any $\beta \in \mathbb{Z}_+^{2n+1}$ put $|\beta| = 2\beta_0 + \beta_1 + \cdots + \beta_n$.

Define $\Psi^m(\mathbb{H}_n)$ as

$$\Psi^m(\mathbb{H}_n) = \{ f \in C^\infty(\mathbb{R}^{4n+2}) : \forall \alpha, \beta \in \mathbb{Z}_+^{2n+1}, \exists \text{ const. } C_{\alpha,\beta}$$

such that $\left| \left(\frac{\partial}{\partial x}\right)^\alpha \left(\frac{\partial}{\partial y}\right)^\beta f(x,y) \right| \leq C_{\alpha,\beta} (1 + [y])^{m - [\beta]} \}$.

We say that such f are symbols of order m

Let

$$O_p \Psi^m = \{ f(\hat{x}, \hat{X}) : f \in \Psi^m(\mathbb{H}_n) \}.$$

Indicate the main formulas of the Symbolic Calculus:

(I) If $f \in \Psi^m(\mathbb{H}_n)$ then $f(\hat{x}, \hat{X})^* = \bar{f}(\hat{x}, \hat{X}) \in O_p \Psi^m$;

(II) Let $f_j \in O_p \Psi^{m_j}(\mathbb{H}_n), j = 1,2$. Then there exists $f \in \Psi^{m_1 + m_2}(\mathbb{H}_n)$

such that

$$f(\hat{x}, \hat{X}) - f_1(\hat{x}, \hat{X}) f_2(\hat{x}, \hat{X}) \in \Psi^{m_1 + m_2 - 1}(\mathbb{H}_n)$$

and

$$(f(\hat{x}, \hat{X}) u)(x) = [f_1(x, \hat{X}) f_2(x, (\hat{X})) u](x);$$

(III) Assume that $f \in \Psi^m(\mathbb{H}_n)$ has the compact support with respect to x. Consider a diffeomorfism χ of a neighborhood of the support. Assume that χ conserves the contact form ω up to a functional multiplier (i. e. χ is a contact mapping).

Let

$$f^\chi(x, y) = f(\chi(x), [d\chi(x)]^* y).$$

Then

$$\chi^{*} f(\hat{x}, \hat{X}) \chi^{*-1} \in O_p \Psi^m$$

and

$$f^{\chi}(\hat{x}, \hat{X}) - \chi^{*} f(\hat{x}, \hat{X}) \chi^{*-1} \in O_p \Psi^{m-1}$$

The last property permits the standard extension of $O_p\Psi^m$ to any contact manifold M. Thus we have $O_p \Psi^m(M)$ with a symbolic Calculus as above.

Now we introduce a subclass $\Psi_0^m(H_n)$ of $f \in \Psi^m(H_n)$ such that for every f there exists a $f_0 \in \Psi^m(H_n)$ such that

(i) $f_0(x, \delta_\tau X) = t^m f(x, X), \quad \forall t > 1, \quad \forall |X| > 1.$

(ii) $f - f_0 \in \Psi^{m-1}(H_n)$

We say that the symbols f have principal parts f_0.

The formula (II) of the Symbolic Calculus shows that in principal parts the product of pseudo-differential operators is the product of left-invariant operators depending on x as a parameter. For study of such products it is very convenient to use the Heisenberg-Fourier transform on the H_n. It is defined by means of the non-degenerate series of unitary represen - tations of the group. By the Stone -von Neumann theorem they are equiva- lent to the representations π_μ depending on non -zero real parameter μ in the space $\mathcal{L}^2(\mathbb{R}^n_\tau)$ such that

$$\pi_\mu(X_0) = \mu 1, \quad \pi_\mu(X') = 2\mu \hat{\tau}, \quad \pi_\mu(X'') = 2i \frac{\partial}{\partial \tau}.$$

For $\varphi \in \mathcal{S}(H_n)$ the Fourier-Heisenberg transform is by definition

$$\pi_\mu(\varphi) = \int_{H_n} \pi_\mu(x) \varphi(x)\, dx, \qquad (\mu \neq 0)$$

This is an integral operator with Schwartz kernel

$$(***)\quad \mathcal{K}_\varphi(\tau,\vartheta,\mu)=\int_{\mathbb{H}_n} e^{i\mu[x_0+(\tau+\vartheta)x']}\varphi(x_0,x',\frac{\tau-\vartheta}{2})\,dx_0\,dx'$$

so that

$$\pi_\mu(\varphi)\;:\;\mathcal{S}'(\mathbb{H}_n)\to\mathcal{S}(\mathbb{H}_n)$$

We see that this formula (***) is valid for any distribution from $\mathcal{S}'(\mathbb{H}_n)$ so we can extend the definition of π_μ to all $\mathcal{S}'(\mathbb{H}_n)$.

This leads to a representation of left-invariant operators $f(\hat{X})\in O_p\Psi_0^m$ by operators.

$$\pi_\mu(f(\hat{X}))=f(\mu 1,2\mu\hat{\tau},2i\frac{\partial}{\partial\tau})\in O_p W_0^m .$$

Moreover

$$\pi_\mu(f_1(\hat{X})f_2(\hat{X}))=\pi_\mu(f_1(\hat{X}))\,\pi_\mu(f_2(\hat{X}))$$

and the principal part of $\pi_\mu(f(\hat{X}))$ is given by means of the principal part of f $(\hat{X})$:

$$(\pi_\mu(f(\hat{X})))_0 = f_0\,(0,2\mu\hat{\tau},2i\frac{\partial}{\partial\tau})$$

Let $f_{oo}(X)$ be the δ_τ-homogeneous function on the $\mathbb{R}^{2n+1}$ which coincide with $f_o(X)$ far from origin . Consider the operators $\pi_\mu(f_{oo}(\hat{X}))$ By δ_t-homogenity we have

$$\pi_\mu(f_{oo}(\hat{X}))=|\mu|^{-m/2}V_{\sqrt{|\mu|}}\,\pi_{\operatorname{sgn}\mu}\,V^{-1}_{\sqrt{|\mu|}}\qquad\text{where } V_t u(\tau)=u(\frac{\tau}{t})$$

Therefore there are significant only $\mu=\pm 1$

Finally we define a σ_m-symbol of an operator $f(\hat{x},\hat{X})\in O_p\Psi_0^m$ as an operator valued function on the manifold of contact directions

$$\sigma_m(f(\hat{x},\hat{X}))_{\omega(x)} = f_{oo}(x,1,2\hat{\tau},2i\frac{\partial}{\partial\tau}),$$

$$\sigma_m(f(\hat{x},\hat{X}))_{-\omega(x)} = f_{oo}(x,-1-2\hat{\tau},\ 2i\frac{\partial}{\partial\tau})$$

$$\sigma_m(f(\hat{x},\hat{X}))_{\pm\omega(x)} \in O_p W_o^m$$

The σ_m-symbol reveals usual properties:

(i) if $f_1 \in \psi^{m_1}(\mathbb{H}_n)$, $f_2 \in \psi^{m_2}(\mathbb{H}_n)$,

then $\sigma_{m_1+m_2}(f_1(\hat{x},\hat{X}), f_2(\hat{x},\hat{X})) = \sigma_{m_1}(f_1(\hat{x},\hat{X}))\sigma_{m_2}(f_2(\hat{x},\hat{X}))$

(ii) $\sigma_m(f(\hat{x},\hat{X})^*) = \sigma_m(f(\hat{x},\hat{X}))^*$

(iii) $\sigma_m(f(\hat{x},\hat{X})) = 0$ if and only if $f(\hat{x}\ \hat{X}) \in O_p\psi^{m-1}$

(iiii) the σ_m-symbol belongs to the contact structure so it can be transfered to any $f(\hat{x},\hat{X}) \in O_p\psi^m(M)$.

Examples: Everywhere M is a strongly pseudoconvex boundary in $\mathbb{C}^{n+1}$.

(a) Let $\Box_M$ be the Kohn sub - Laplacian on the space of (o, q) -forms. Then

$$\Box_M \in O_p\,\psi^2(M) \quad \text{and} \quad \sigma_2(\Box_M) = \left(-\frac{\partial^2}{\partial\tau^2} + \tau^2 + n - 2q\right)1_{T^{0,q}}$$

The operator $-\frac{\partial^2}{\partial\tau^2} + \tau^2$ is the energy operator of the harmonic oscillator of Quantum Mechanics. Actually this example has appeared in /4/ and by the way it served an origin of our study.

(b) The induced Cauchy-Riemann operator $\bar{\partial}_M$ on the (o, q) -forms belongs to $O_p\psi^1(M)$ with

$$\sigma_1(\bar{\partial}_M)_{\pm\omega(x)} = \left(-\frac{\partial}{\partial\tau} \pm \hat{\tau}\right)1_{T^{oq}}$$

(c) The Cauchy-Henkin integral can be considered as operators S on M if we take their boundary values; they belong to $O_p\psi^o(M)$

and their symbol is

$\sigma_1(S)_{\omega(x)}$ = orthogonal projector on the linear span of $e^{-\tau^2/2}$

$\sigma_1(S)_{-\omega(x)}$ = zero

We say that an operator $f(\hat{x}, \hat{X}) \in O_p\Psi^m$ is σ-elliptic if

(Ell-1) $\sigma_m(f(\hat{x}, \hat{X}))_{\pm\omega(x)} \in Ell\, W_0^m$

(Ell-2) The operators $\sigma_m(f(\hat{x}, \hat{X}))_{\pm\omega(x)}$ are invertible in $\mathcal{S}(\mathbb{R}^n_\tau)$

Let M be a compact contact manifold. We can introduce a scale of anisotropic spaces of functions and distributions on M

$$\mathcal{S}^k_p(M) \quad , \quad 1 < p < \infty \; , \quad -\infty < k < +\infty$$

of E. Stein (cf. /4/).

If $f(\hat{x}, \hat{X}) \in O_p\Psi^m$ then $f(\hat{x}, \hat{X})$ is bounded from $\mathcal{S}^k_p(M)$ to $\mathcal{S}^{k-m}_p(M)$ for any k

The following properties are equivalent for $f(\hat{x}, \hat{X}) \in O_p\Psi^m$:

a) $f(\hat{x}, \hat{X})$ is a σ-elliptic operator.

b) The apriori estimate

$$\|u\|_{k+m} \leq Const\,(\|f(\hat{x}, \hat{X})u\|_k + \|u\|_{k'})$$

is valid in Stein norms for a (and therefore for any) $k \in \mathbb{R}$, $k' < k$

c) $f(\hat{x}, \hat{X})$ is a Fredholm operator from $\mathcal{S}^k_p(M)$ to $\mathcal{S}^{k-m}_p(M)$ for a (and therefore for any) $k \in \mathbb{R}$.

Remark

CENTRO INTERNAZIONALE MATEMATICO ESTIVO

(C.I.M.E.)

AN INDEX FORMULA FOR ELLIPTIC BOUNDARY PROBLEMS

A. DYNIN

Corso tenuto a Bressanone dal 16 al 24 giugno 1977

I give an analytical formula for index of elliptic boundary problems for scalar differential operator and for some systems of differential operators of even order in bounded domains with smooth boundaries in euclidean space.

1. Notation.

$x = (x_1, \ldots, x_n) \in \mathbb{R}^n$, $\xi = (\xi_1, \ldots, \xi_n) \in (\mathbb{R}^n)^*$,

$D_j = \frac{1}{i}\frac{\partial}{\partial x_j}$, $D = (D_1, \ldots, D_n)$, $\alpha = (\alpha_1, \ldots, \alpha_n) \in \mathbb{Z}^n_+$,

$D^\alpha = D_1^{\alpha_1} \ldots D_n^{\alpha_n}$, $\xi^\alpha = \xi_1^{\alpha_1} \ldots \xi_n^{\alpha_n}$, $|\alpha| = \alpha_1 + \ldots + \alpha_n$.

The term "smooth" always means C^∞.

Let U be an open bounded domain in $\mathbb{R}^n$ with smooth boundary Y. Points of Y are denoted y. Cotangent vectors at y with length 1 are denoted τ_y. The $(2n-3)$- manifold $S(Y)$ of all such τ_y is supplied with canonical orientation: the manifold $T^*(Y)$ of all cotangent vectors of Y is $\mathbb{R}^{n-1} \times (\mathbb{R}^{n-1})^*$ locally. Let $(y_1, \ldots, y_{n-1})$ be any system of coordinates on $(\mathbb{R}^{n-1}$ and $(\eta_1, \ldots, \eta_{n-1})$ the dual system of coordinates on $(\mathbb{R}^{n-1})^*$. Then the orienting $(2n-2)$- form $\Omega = (dy_1 \wedge d\eta_1) \wedge \ldots \wedge (dy_n \wedge d\eta_n)$ does not depend on the choice of le local coordinates and therefore gives an orientation of $(\mathbb{R}^{n-1})^*$. Now let $\tau : T^*(Y) \to \mathbb{R}$ be the euclidean metric, so that $S(Y) = \tau^{-1}(1)$. Then the orienting (2n-3)- form ω on the $S(Y)$ is defined by its property

$d\tau \wedge \omega = \Omega$.

2. Elliptic Boundary Problems.

Let A be a scalar differential operator of order $2m$ with smooth coefficients

$$A : u(x) \to \sum_{|\alpha| < 2m} a_\alpha(x) D^\alpha u(x),\ u \in C^\infty(\bar{U})$$

Its principal symbol is the function on $T^*(\bar{U}) = \bar{U} \times (\mathbb{R}^n)^*$

$$\sigma(A)(x, \xi) = \sum_{|\alpha| = 2m} a_\alpha(x)$$

The operator A is assumed elliptic i.e.

$$\sigma(A)(x, \xi) \neq 0,\ \forall x \in \bar{U},\ \forall \xi \in (\mathbb{R}^n)^*$$

Consider λ -polynomial with coefficients from $C^\infty(S(Y))$ of order $2m$

$$\hat{\sigma}(A)(\tau_y, \lambda) = \sigma_A(y, \tau_y + \lambda \nu_y)$$

where ν_y is the inward unit conormal at y.

For each τ_y this polynomial has no root with zero imaginary part.

If $n > 2$ there are exactly m roots with positive imaginary part (this is an easy consequence of the connectivity of $S(Y)$). If $n = 2$ then we assume this property especially.

We can factorize the λ -polynomial into the product of two polynomials with smooth coefficients

$$\hat{\sigma}(A)(\tau_y, \lambda) = \sigma^+(\tau_y, \lambda)\, \sigma^-(\tau_y, \lambda),$$

where all roots of $\sigma^+(\tau_y, \lambda)$ are in the upper complex λ - halfplane and all roots of $\sigma^-(\tau_y, \lambda)$ are in the lower λ - halfplane.

Consider a boundary problem

$$(a)\begin{cases} Au = f \quad , f \in C^{\infty}(\bar{U}) \\ B_j u = g_j \quad g_j \in C^{\infty}(Y), \quad j = 1, \ldots, m \end{cases}$$

Here B_j are boundary differential operators of order m_j with smooth coefficients

$$B_j : u \to \sum_{|\alpha| < m_j} b_\alpha(y)\, D^\alpha u / Y,$$

$$B_j : C^{\infty}(\bar{U}) \to C^{\infty}(Y).$$

We suppose that the boundary problem satisfies the Shapiro-Lopatinsky condition of ellipticity which we take in the Agmon-Douglis-Nirenberg version /1/ (cf. lectures by F. Trèves):

Consider λ-polynomials of degree m_j with smooth coeffients on $S(Y)$

$$\hat{\sigma}(B_j)(\tau_y, \lambda) = \sigma(B_j)(y, \tau_y + \lambda \nu_y) \equiv$$

$$\equiv \sum_{|\alpha| = m_j} b_{j\alpha}(y)\, (\tau_y + \lambda \nu_y)^\alpha, \quad j = 1, \ldots, m.$$

The Agmon-Douglis-Nirenberg condition is the linear independence of these polynomials modulo λ-polynomial $\sigma^+(\tau_y, \lambda)$ for every τ_y.

We can represent this condition in an equivalent form. Let $r_j(\tau_y, \lambda)$ be the remainder from division of λ-polynomial $\hat{\sigma}(B_j)(\tau_y, \lambda)$ by λ-polynomial $\sigma^+(\tau_y, \lambda)$:

$$\hat{\sigma}(B_j)(\tau_y, \lambda) = q_j(\tau_y, \lambda)\, \sigma^+(\tau_y, \lambda) + r_j(\tau_y, \lambda)$$

where q_j and r_j are λ-polynomials and the degree of $r_j(\tau_y, \lambda)$ is less than m. Let

$$\tau_j(\tau_y, \lambda) = \sum_{k=0}^{m-1} \tau_{kj}(\tau_y)\lambda^k, \quad j = 1, \ldots, m$$

Consider the square (m x m) - matrix valued function on S (Y)

$$\tau(\mathfrak{a}) = (\tau_{kj})$$

The Agmon-Douglis-Nirenberg condition is obviously equivalent to non-degeneracy condition

(ADN) $\det \tau(\mathfrak{a})(\tau_y) \neq 0, \quad \forall \tau_y \in S(Y)$

3. The Index Formula

As usual the elliptic boundary problem ($\mathfrak{a}$) leads to a linear continuous operator

$$\mathfrak{a} = (A, B_1, \ldots, B_m) : C^\infty(\bar{U}) \to C^\infty(\bar{U}) \times (C^\infty(Y))^m$$

which is a Fredholm operator and therefore has a finite index

$$\operatorname{ind} \mathfrak{a} = \dim \operatorname{Ker} \mathfrak{a} - \dim \operatorname{coker} \mathfrak{a}$$

(see e. g. / 2 / and / 8 /).

It is known that the index depends only on the symbol

$$\sigma(\mathfrak{a}) = (\sigma(A), \sigma(B_1), \ldots, \sigma(B_m)).$$

We express it by means of the $\tau(\mathfrak{a})$ which is defined by $\sigma(\mathfrak{a})$

(1) $$\operatorname{ind} \mathfrak{a} = \frac{(-1)^n}{(2\pi i)^{n-1}} \frac{(n-2)!}{(2n-3)!} \int_{S(Y)} \operatorname{Sp} \left[\tau(\mathfrak{a})^{-1} d\tau(\mathfrak{a})\right]^{2n-3}$$

The intengrand is the trace of (2n-3)-power of (m x m) -matrix valued differential 1-form $\tau(\mathfrak{a})^{-1} \, d\tau(\mathfrak{a})$ in the exterior algebra of matrix valued differential forms. So we integrate (2n-3)-form over the (oriented) manifold S (Y).

In particular the Index Formila shows that if $m < n-1$ then ind $\mathcal{A} = 0$

The prof. of (1) involves a special homotopy of $\sigma(\mathcal{A})$ in the space of symbols of elliptic boundary problems for the A with pseudo-differential boundary operators. (By the way, this is the first place where pseudo-differential operators of positive order were introduced as early as in 1961: see / 4 / and / 5 /.)

The homotopy is

$$\hat{\sigma}(t)(B_j)(\tau_y,\lambda) = (1-t)\, q_j(\tau_y,\lambda)\, \sigma^+(\tau_y,\lambda) + \tau_j(\tau_y,\lambda)$$

Here $0 \leq t \leq 1$ and

$$\hat{\sigma}_{(0)}(B_j)(\tau_y,\lambda) = \hat{\sigma}(B_j)(\tau_y,\lambda),$$

$$\hat{\sigma}_{(1)}(B_j)(\tau_y,\lambda) = \tau_j(\tau_y,\lambda).$$

This homotopy may be covered by homotopy of boundary value problems ($\mathcal{A}_t$) for the same operator A which all are elliptic by $\tau(\mathcal{A}_t) = \tau(\mathcal{A})$ and the condition (ADN) is satisfied (cf. /5/ and lectures by F. Treves).

Stability os Index under homotopies implies

(2) ind $\mathcal{A}$ = ind $\mathcal{A}_1$.

The (1x(m+1) -matrix $\sigma(\mathcal{A}_1)$ can be factorized

(3) $\sigma(\mathcal{A}_1) = \sigma(\mathcal{D})\,(1 \oplus \tau(\mathcal{A}))$.

where

$$\sigma(\mathcal{D}) = (\sigma(A), 1, \nu, \ldots ; \nu^{m-1})$$

is the symbol of the (elliptic) Dirichlet problem for the operator A.

We consider $\tau(\mathcal{A})$ as the symbol $\sigma(R_{\mathcal{A}})$ of a system of pseudodifferential operators $R_{\mathfrak{a}}$ in $(C^{\infty}(Y))^{m}$ which is elliptic by (ADN). (Strictly speaking the $R_{\mathfrak{a}}$ is elliptic in the Douglis-Nirenberg sense only, otherwise we have to modify the Dirichlet problem, cf. /5/ and lectures by F. Treves.)

Now by algebraic properties of Index the equality (3) implies

$$\operatorname{ind} \mathfrak{a}_1 = \operatorname{ind} D + \operatorname{ind} (1 + R_{\mathfrak{a}}) = \operatorname{ind} D + \operatorname{ind} R_{\mathfrak{a}}$$

But $\operatorname{ind} D = 0$ (cf. /2/). Therefore (by (2))

$$\operatorname{ind} \mathfrak{a} = \operatorname{ind} R_{\mathfrak{a}}$$

Finally the Index Formula (1) coincides with the Index Formula discovered by A. Dynin and B. Fedosov (cf. /6/) for the elliptic pseudodifferential system $R_{\mathfrak{a}}$ on the manifold without boundary Y. Of course such formula can be derived from the famous Atiyah - Singer formula and actually this was accomplished by the author (Proceedings of the Conference on the Mathematical Methods in Physics, Dubna, 1964) and by B. Fedosov /6/. Nowadays B. Fedosov /7/ has found a completely analytical proof of the formula. Therefore we have an elementary proof of our formula (1).

4. The Index Formula for Systems.

Consider now an elliptic (NxN)-system A of order $2m$. Suppose that we can again factorize its principal symbol

$$\hat{\sigma}(A)(\tau_y, \lambda) = \sigma^{+}(\tau_y, \lambda)\, \sigma^{-}(\tau_y, \lambda), \quad \deg \sigma^{\pm} = m.$$

By a theorem of Lopatinsky this factorization exists if and only if the rank of the (NxmN) -matrix

$$\oint_{\gamma_+} \hat{\sigma}(\tau_y, \lambda)^{-1}(1_N, \lambda 1_N, \dots, \lambda^{m-1} 1_N)\, d\lambda$$

is maximal for any τ_y and any curve γ_+ embracing the spectra of the $\hat{\sigma}(\tau_y, \lambda)$ in the upper complex λ-halfplane (cf. /2/).

We may reproduce all above considerations for boundary problem $\mathcal{a} = (A, B_1, \dots, B_m)$ where B_j are now (NxN) -systems of boundary differential operators. The (ADN)-condition means that the corresponding (mN x mN)-matrix $\tau(\mathcal{a})$ is non-singular for any $\tau_y \in S(Y)$

As above we have

(4) $$\operatorname{ind} \mathcal{a} = \operatorname{ind} \mathcal{D} + \operatorname{ind} R_{\mathcal{a}}$$

where

$$D = \left(A, 1_N, \frac{1}{i}\frac{\partial}{\partial \nu} 1_N, \dots, \left(\frac{1}{i}\frac{\partial}{\partial \nu}\right)^{m-1} 1_N\right)$$

is the operator of the Dirichlet problem for the elliptic system A . However its index can be non-trivial.

In /2/ M. Agranovich proved that the ind $\mathcal{D}$ is equal to the index of an elliptic system of pseudo-differential operators on a manifold without boundary . Applying tha analytic Index formula to the Agranovich system we can derive the following result:

Consider the space of cotangent balls of the $\bar{U}$

$$D(\bar{U}) \simeq \bar{U} \times D, \quad D = \{\xi ;\ |\xi| \leq 1\}.$$

The boundary of $D(\bar{U})$ is

$$\partial D(\bar{U}) \simeq (Y \times D) \cup S(\bar{U}), \quad S(\bar{U}) \simeq U \times \partial D$$

The symbol $\sigma(A)$ is defined on the $S(\bar{U})$. Extend it artificially on the rest $Y \times D$. Note that $Y \times D$ has a representation

$$Y \times D \simeq \left\{ \frac{t}{\sqrt{1+\lambda^2}} (\tau_y + \lambda \nu_y) : \tau_y \in S(Y),\ 0 \leq t \leq 1,\ -\infty \leq \lambda \leq \infty \right\}$$

Let $\tilde{\sigma}(A) \mid S(\bar{U}) = \sigma(A) \mid S(\bar{U})$,

$$\tilde{\sigma}(A) \left(\frac{t}{\sqrt{1+\lambda^2}} (\tau_y + \lambda \nu_y) \right) =$$

$$(1+\lambda^2)^{-m} \left[t \overset{+}{\sigma}(\tau_y, \lambda) + (1-t)(\lambda - i)^m 1_N \right] \left[t \sigma^-(\tau_y, \lambda) + (1-t)(\lambda+i)^m 1_N \right]$$

This is the desired extension of $\sigma(A)$ to a mapping $\tilde{\sigma}(A)$ of all $\beta D(\bar{U})$ into $GL(N, \mathbb{C})$

Now if $\beta D(\bar{U})$ is supplied with canonie orientation then

$$\text{ind}\, \mathcal{D} = \frac{(-1)^{n+1}}{(2\pi i)^n} \frac{(n-1)!}{(2n-1)!} \int_{\beta D(\bar{U})} \text{Sp} \left[\tilde{\sigma}(A)^{-1} d\tilde{\sigma}(A) \right]^{2n-1}$$

Note that if $N < n$ then $\text{ind}\, \mathcal{D} = 0$.

Thus we obtained the Index Formula

$$\text{ind}\, \mathcal{a} =$$

$$\frac{(-1)^{n+1}}{(2\pi i)^n} \frac{(n-1)!}{(2n-1)!} \int_{\beta D(\bar{U})} \text{Sp} \left[\tilde{\sigma}(A)^{-1} d\tilde{\sigma}(A) \right]^{2n-1} +$$

$$\frac{(-1)^n}{(2\pi i)^{n-1}} \frac{(n-2)!}{(2n-3)!} \int_{S(Y)} \text{Sp} \left[\tau(a)^{-1} d\tau(a) \right]^{2n-3}$$

At last there exists the general Index Formula of Atiyah and Bott. It was proved by the author (unpublished) and Boutet de Monvel /3/.

<u>References</u>

/1/ Agmon, S., Douglis, A., Nirenberg, L., Communications on Pure and Applied Mathematics, v. 12 (1959), 623 - 727.

/2/ Agranovich, M. S., Uspekhi, v. 20 (1965), no. 5, 3-120.

/3/ Boutet de Monvel, Acta Math. , v. 126 (1971), no. 1-2, 11-51.

/4/ Dynin, A.S., Donlady V. 141 (1961), 21-23.

/5/ Dynin, A. S., Doklady, V. 141 (1961), 285 - 287.

/6/ Fedosov, B. V., Funkzionalny Analiz i ego Prilojeniya, v, 4 (1970) no 4, 83 -84.

/7/ Fedosov, B, V., Trudy, v. 30 (1974), 159-241

/8/ Hormander, L., Linear Partial Differential Operators, Berlin, 1963.

Of course every σ-elliptic operator is hypoelliptic. It is interesting that hypoelliptic δ_τ-homogeneous left-invariant operators of Rockland and Beals (/1/, /6/) are σ-elliptic.

References

/1/ Beals, R. , Seminaire Goulaic-Schwartz, exp. XIX, 1976-1977

/2/ Dynin, A., Soviet Math. Doklady, v. 16 (1975) No 6, 1608-1612.

/3/ Dynin, A. Soviet Math. Doklady, v. 17 (1976) No 2, 5o8 - 513.

/4/ Folland, G. B. , Stein E. M. Communications on Pure and Appl. Math. V. 27 (1974) , n. 4; 429 - 522

/5/ Shubin, M. A. Soviet Math. Doklady, v. 12 (1971), No 1, 147 - 151

/6/ Rockland, Ch, Hypoellipticity on the Heisenberg group, Preprint

CENTRO INTERNAZIONALE MATEMATICO ESTIVO

(C.I.M.E.)

GENERAL MIXED BOUNDARY PROBLEMS FOR ELLIPTIC

DIFFERENTIAL EQUATIONS

G. ESKIN

Hebrew University of Jerusalem

Corso tenuto a Bressanone dal 16 al 24 giungo 1977

General mixed boundary problems for elliptic differential equations

Gregory Eskin
Hebrew University of Jerusalem

0. Introduction

Let G be a bounded domain in R^n with a smooth $n-1$-dimensional boundary Γ. Assume that Γ is divided by a smooth $n-2$ dimensional manifold Γ_0 on two parts Γ_1 and Γ_2, so that $\bar{\Gamma}_1 \cup \bar{\Gamma}_2 = \Gamma$, $\bar{\Gamma}_1 \cap \bar{\Gamma}_2 = \Gamma_0$. Consider in G a second order elliptic equation

$$(0.1) \quad L(x,D)u = \sum_{i,k=1}^{n} a_{ik}(x) \frac{\partial^2 u}{\partial x_i \partial x_k} + \sum_{k=1}^{n} b_k(x) \frac{\partial u}{\partial x_k} + c(x)u = f(x),$$

with a mixed boundary conditions

$$(0.2) \quad B_1(x,D)u \big|_{\Gamma_1} = g_1,$$

$$(0.3) \quad B_2(x,D)u \big|_{\Gamma_1} = g_2,$$

where $B_k(x,D)$ are differential operators of order m_k, $k = 1,2$. A classical example of a mixed elliptic boundary problem is the following

$$(0.4) \quad \Delta u = f(x), \quad x \in G,$$

$$(0.5) \quad u \big|_{\Gamma_1} = g_1, \quad \frac{\partial u}{\partial n} \Big|_{\Gamma_2} = g_2,$$

where $\frac{\partial}{\partial n}$ is the normal derivative. Some other examples of mixed problems will be given later.

The mixed problem for elliptic equations of second order was considered by Lienard [16], Fichera [11], Miranda [18], Magenes and Stampaccia [17], Stampaccia [25]. The general case of mixed problems for higher order elliptic equations with two independent variables was considered by J. Peetre [20]. The L_p-theory of such problems was given by E. Shamir [21].

In 1962 the author has studied a general mixed problem for a multidimensional equation of the second order. These investigations were continued in a series of joint papers with M. Višik, where a general theory of pseudodifferential equations on a manifold with a boundary was developed.

The present lectures are based on the mongraph [6] where this theory was completed and simplified (see also paper [5]) In [5] and [6] one can find also more detailed references.

As usually in the theory of the elliptic boundary problems we shall consider at first the case when $G = \mathbb{R}^n_+ = \{(x',x_n), x_n > 0\}$, $x' = (x'',x_{n-1})$, $\Gamma_1 = \mathbb{R}^{n-1}_+ = \{(x'',x_{n-1}), x_{n-1} > 0\}$, $\Gamma_2 = \mathbb{R}^{n-1}_- = \{(x'',x_{n-1}), x_{n-1} < 0\}$, $\Gamma_0 = \bar{\Gamma}_1 \cap \bar{\Gamma}_2 = \mathbb{R}^{n-2}$ and L, B_1, B_2 are homogeneous differential operators with constant coefficients:

$$(0.6) \qquad L(D)u(x) = f(x), \quad x \in \mathbb{R}^n_+ ,$$

$$(0.7) \qquad B_1(D)u(x)\big|_{\mathbb{R}^{n-1}_+} = g_1(x'), \quad B_2(D)\, u(x)\big|_{\mathbb{R}^{n-1}_-} = g_2(x') .$$

The study of the problem (0.6), (0.7) is not convenient, because the polynomials $L(\xi)$, $B_1(\xi)$, $B_2(\xi)$ are vanished for $\xi = 0$.

Denote

$$\hat{L}(\xi'',\xi_{n-1},\xi_n) = L(1 + |\xi''|)\omega,\xi_{n-1},\xi_n), \quad \omega = \frac{\xi''}{|\xi''|} .$$

Then $\hat{L}(\xi)$ will be polynomial on ξ_{n-1} and ξ_n and the order of $\hat{L}-L$

will be less than the order of $L(\xi)$. Since $L(\xi)$ is elliptic, i.e. $L(\xi) \neq 0$ for $\xi \neq 0$ we have $C_1(1 + |\xi|)^2 \leq |\hat{L}(\xi)| \leq C(1 + |\xi|)^2$. Analogously we set

$$\hat{B}_i(\xi) = B_i((1 + |\xi''|)\omega, \ \xi_{n-1}, \xi_n), \ i = 1,2.$$

Instead of the problem (0.6),(0.7) it is more convenient to consider the following problem

$$(0.8) \qquad \hat{L}(D)\, u(x) = f(x), \quad x \in \mathbb{R}^n_+,$$

$$(0.9) \qquad \hat{B}_1(D)\, u(x) \big|_{\mathbb{R}^{n-1}_+} = g_1(x'), \ \hat{B}_2(D) u(x) \big|_{\mathbb{R}^{n-1}_-} = g_2(x') \, ;$$

where $\hat{L}(D)$, $\hat{B}_1(D)$, $\hat{B}_2(D)$ are pseudodifferential operators with symbols $\hat{L}(\xi)$, $\hat{B}_1(\xi)$, $\hat{B}_2(\xi)$. We note that the problem (0.8), (0.9) differ from (0.6), (0.7) only by the terms of lower order. Therefore the conditions of the normal solvability for the elliptic problem (0.1), (0.2), (0.3) which depends only on the principal parts of $L(x,D)$, $B_i(x,D)$, $i = 1,2$, will be the same as the conditions of the solvability and the uniqueness for the problem (0.8), (0.9).

1. Sobolev's spaces with weights

Even for infinitely differentiable $f(x)$, $g_1(x)$, $g_2(x)$ the solution $u(x)$ of the problem (0.8), (0.9) has, in general, singularities for $x_n = x_{n-1} = 0$. Outside of the hyperplane $x_n = x_{n-1} = 0$ the solution $u(x)$ is a C^∞-function. Therefore it is natural to study the problem (0.8), (0.9) in the functional spaces with the weights which vanish for $x_n = x_{n-1} = 0$. Let $s \in \mathbb{R}$ be arbitrary. The Sobolev space $H_s(\mathbb{R}^n)$ is a space of tempered distributions $u(x)$ with the norm

$$\|u\|_s^2 = \int_{-\infty}^{\infty} (1 + |\xi|^2)^s \, |\tilde{u}(\xi)|^2 \, d\xi \quad ,$$

where $\tilde{u}(\xi) = \int_{-\infty}^{\infty} u(x)\, e^{i(x,\xi)} dx$ is the Fourier transform of the distribution $u(x)$

As usually we define the space $H_s(\mathbb{R}^n_+)$ as the restriction of $H_s(\mathbb{R}^n)$ to $\mathbb{R}^n_+$ with the following norm

$$\|u\|_s^+ = \inf_{\ell u \in H_s(\mathbb{R}^n)} \|\ell u\|_s ,$$

where $\ell u(x)$ is an arbitrary function in $H_s(\mathbb{R}^n)$ which restriction to $\mathbb{R}^n_+$ is equal to $u(x)$. Let $N \geq 0$ be an integer. By $H_{s,N}(\mathbb{R}^n)$ we denote the space of tempered distributions $u(x)$ with the following finite norm:

$$\|u\|^2_{s,N} = \sum_{k_1+k_2=0}^{N} \|x_n^{k_1} x_{n-1}^{k_2} u\|^2_{s+k_1+k_2}$$

The space $H_{s,N}(\mathbb{R}^n_+)$ is defined in the same way as $H_s(\mathbb{R}^n_+)$. We note that $H_{s,0}(\mathbb{R}^n) = H_s(\mathbb{R}^n)$, $H_{s,0}(\mathbb{R}^n_+) = H_s(\mathbb{R}^n_+)$. Denote by $H'_s(\mathbb{R}^{n-1})$ the Sobolev space in $\mathbb{R}^{n-1}$. We shall denote the norm in $H'_s(\mathbb{R}^{n-1})$ by $\|v\|'_s$, so that

$$(\| v(x')\|'_s)^2 = \int_{-\infty}^{\infty} (1 + |\xi'|^2)^s \,|\tilde{v}(\xi')|^2 \, d\xi'$$

Let $H'_{s,N}(\mathbb{R}^{n-1})$ be the space of tempered distributions with the finite norm

$$(\|v\|'_{s,N})^2 = \sum_{k=0}^{N} (\|x^k_{n-1} v\|'_{s+k})^2$$

The spaces $H'_s(\mathbb{R}^{n-1}_{\pm})$, $H'_{s,N}(\mathbb{R}^{n-1}_{\pm})$ are defined in the same way as $H_s(\mathbb{R}^n_+)$, $H_{s,N}(\mathbb{R}^n_+)$.

We note some simple properties of the spaces $H_{s,N}$ and $H'_{s,N}$.

<u>Lemma 1.1:</u> <u>Let</u> $\left|\frac{\partial^k A(\xi')}{\partial \xi^k_{n-1}}\right| \leq C_k(1 + |\xi'|)^{\alpha-k}$, $\forall k \geq 0$ <u>and let</u> $A(D')$ <u>be a pseudodifferential operator</u> (ψ.d.O.) <u>with the symbol</u> $A(\xi')$, <u>i.e.</u>

$$A(D)v = \frac{1}{(2\pi)^{n-1}} \int_{-\infty}^{\infty} A(\xi')\tilde{v}(\xi')e^{-i(x',\xi')}d\xi' \text{ for } v \in C_0^\infty(\mathbb{R}^{n-1})$$

Then for arbitrary s and N

$$\|A(D')v\|'_{s,N} \leq C\|v\|'_{s+\alpha,N}$$

Analogously if $\left|\frac{\partial^{k_1+k_2} B(\xi)}{\partial\xi_{n-1}^{k_1}\partial\xi_n^{k_2}}\right| \leq C(1+|\xi|)^{\alpha-k_1-k_2}, \forall\, k_1 \geq 0,\ k_2 \geq 0$

then $\|B(D)u\|_{s,N} \leq C\|u\|_{s+\alpha,N}$

The proof of Lemma 1.1 follows immediately from the fact that the norm in $H'_{s,N}$ is equivalent to the following norm

$$(\|v\|'_{s,N})^2 \approx \sum_{k=0}^{N} \int_{-\infty}^{\infty} (1+|\xi'|^2)^{s+k} \left| \frac{\partial^k}{\partial\xi_{n-1}^k} \tilde{v}(\xi')\right|^2 d\xi'$$

Denote by p' the restriction operator to the hyperplane $x_n = 0$ and by p'_+ (p'_-) the restriction operator to the half-space $x_n = 0$, $x_{n-1} > 0$ ($x_{n-1} < 0$). It follows from the well-known properties of the Sobolev spaces that $\|p'u\|'_{s-\frac{1}{2},N} \leq C\|u\|_{s,N}$ if $u \in H_{s,N}(\mathbb{R}^n)$ and $s > \frac{1}{2}$. The following lemma is less trivial (see [6]):

Lemma 1.2: If $s \leq \frac{1}{2}$, s is noninteger and $s+N > \frac{1}{2}$ then

$$\|p'_\pm u\|'_{s-\frac{1}{2},N} \leq C\|u\|_{s,N}$$

2. Solution of the mixed problem in the half-space.

Consider the mixed problem (0.8), (0.9) assuming that $f \in H_{s-2,N}(\mathbb{R}^n_+)$, $g_1 \in H_{s-m_1-\frac{1}{2},N}(\mathbb{R}^{n-1}_+)$, $g_2 \in H_{s-m_2-\frac{1}{2},N}(\mathbb{R}^{n-1}_-)$. Here s will be chosen later.

We suppose that $s + N > \max_{i=1,2}(m_i + \frac{1}{2})$ and s is noninteger if $s \leq \max_{i=1,2}(m_i+\frac{1}{2})$.

A particular solution of the equation (0.8) in the half-space $\mathbb{R}^n_+$ can be taken in the following form

$$u_0(x) = F^{-1} \frac{\widetilde{\ell f}(\xi)}{\hat{L}(\xi)}$$

where $\ell f \in H_{s-2,N}(\mathbb{R}^n)$ is an arbitrary extension of f to $\mathbb{R}^n$ and F^{-1} is the inverse Fourier transform.

Let $\hat{\lambda}(\xi')$ be the root of equation $\hat{L}(\xi',\xi_n) = 0$ with the negative imaginary part. Then the general solution in $H_{s,N}(\mathbb{R}^n_+)$ of the equation (0.8) has the form

$$(2.1) \qquad u(x) = \frac{1}{(2\pi)^{n-1}} \int e^{-i\, x_n\hat{\lambda}(\xi')-i(x',\xi')}\tilde{v}(\xi')d\xi' + u_0(x) \;,$$

where $v(x')$ is an arbitrary function belonging to $H'_{s-\frac{1}{2},N}(\mathbb{R}^{n-1})$. In order to satisfy the boundary conditions (0.9) we obtain the following system of equations for $v(x')$:

$$(2.2) \qquad \begin{aligned} p'_+\hat{b}_1(D')v(x') &= g_1(x') - p'_+ B_1(D)u_0 \;, \\ p'_-\hat{b}_2(D')v(x') &= g_2(x') - p'_- B_2(D)u_0 \;, \end{aligned}$$

where $b_i(\xi') = B_i(\xi',\lambda(\xi'))$, $\hat{b}_i(\xi') = b_i((1+|\xi'|)\omega,\xi_{n-1})$, $\hat{b}_i(D')$ are ψ.d.o. with the symbols $\hat{b}_i(\xi')$, $i = 1,2$.

We note that $p'_+\hat{B}_1(D)u_0 \in H'_{s-m_1-\frac{1}{2},N}(\mathbb{R}^{n-1}_+)$, $p'_-\hat{B}_2(D)u_0 \in H'_{s-m_2-\frac{1}{2},N}(\mathbb{R}^{n-1}_-)$ (see Lemma 1.2).

We assume that the boundary operators $B_1(D)$ and $B_2(D)$ satisfy the Shapiro-Lopatinskii condition what means that

$$(2.3) \qquad b_i(\xi') \neq 0 \quad \text{for} \quad \xi' \neq 0,\ i = 1,2.$$

Let $\ell g_i(x') \in H'_{s+m_i-\frac{1}{2},N}(\mathbb{R}^{n-1})$ be extensions of g_1 and g_2 to $\mathbb{R}^{n-1}$. Set

$$(2.4) \qquad \begin{aligned} v_-(x') &= \ell g_1 - p'B_1(D)u_0 - \hat{b}_1(D')v \;, \\ v_+(x') &= \ell g_2 - p'B_2(D)u_0 - \hat{b}_2(D')v \;. \end{aligned}$$

Then $v_+ \in H^+_{s-m_2-\frac{1}{2},N}(\mathbb{R}^{n-1})$, $v_- \in H^-_{s-m_1-\frac{1}{2},N}(\mathbb{R}^{n-1})$, where $H^{\pm}_{s,N}$ is the subspace of $H'_{s,N}(\mathbb{R}^{n-1})$ of functions with supports in $\mathbb{R}^{n-1}_{\pm}$.
Taking the Fourier transform in (2.4) we obtain

$$
\begin{aligned}
&\hat{b}_1(\xi')\,\tilde{v}(\xi') + \tilde{v}_-(\xi') = \tilde{h}_1(\xi') \quad , \\
(2.5)\qquad & \\
&\hat{b}_2(\xi')\,\tilde{v}(\xi') + \tilde{v}_+(\xi') = \tilde{h}_2(\xi') \quad ,
\end{aligned}
$$

where $h_i(x') = \ell g_i(x') - p'B_i(D)u_0$, $i = 1,2$.

By exluding $\tilde{v}(\xi')$ from (2.5) we obtain

$$
(2.6)\qquad \tilde{v}(\xi') = -\frac{v_+(\xi')}{\hat{b}_2(\xi')} + \frac{h_2(\xi')}{\hat{b}_2(\xi')} \quad .
$$

So that

$$
(2.7)\qquad \hat{b}_1(\xi')\,\hat{b}_2^{-1}(\xi')\,\tilde{v}_+(\xi') - \tilde{v}_-(\xi') = \hat{b}_1(\xi')\,\hat{b}_2^{-1}(\xi')\tilde{h}_2(\xi') - \tilde{h}_1(\xi') \quad .
$$

Therefore the solution of the equation (2.2) is reduced to the solution of the equation (2.7). The equation (2.7) is called Wiener-Hopf equation or Riemann-Hilbert equation and the way to solve this equation is the factorization of the function $\hat{b}_1(\xi')\hat{b}_2^{-1}(\xi')$.

For the factorization we shall need some properties of the Cauchy integral.

3. The Cauchy integral

Let $f(x') \in C_0^{\infty}(\mathbb{R}^{n-1})$ and let $\tilde{f}(\xi')$ be the Fourier transform of $f(x')$. Denote by $\Pi^{\pm}$ the following operators:

$$
\Pi^+\tilde{f} = \frac{i}{2\pi}\int_{-\infty}^{\infty} \frac{\tilde{f}(\xi'',\eta_{n-1})}{\xi_{n-1}+i0-\eta_{n-1}}\,d\eta_{n-1} = \lim_{\tau\to+0} \frac{i}{2\pi}\int_{-\infty}^{\infty} \frac{\tilde{f}(\xi'',\eta_{n-1})}{\xi_{n-1}+i\tau-\eta_{n-1}}\,d\eta_{n-1} =
$$

$$
= \frac{1}{2}\tilde{f} + \text{v.p.}\,\frac{i}{2\pi}\int_{-\infty}^{\infty} \frac{\tilde{f}(\xi'',\eta_{n-1})}{\xi_{n-1}-\eta_{n-1}}\,d\eta_{n-1} \quad ,
$$

$$\Pi^-\tilde f = -\frac{i}{2\pi}\int_{-\infty}^{\infty}\frac{\tilde f(\xi'',\eta_{n-1})}{\xi_{n-1}-i0-\eta_{n-1}}\,d\eta_{n-1} = \frac{1}{2}\tilde f(\xi') - \text{v.p.}\frac{i}{2\pi}\int_{-\infty}^{\infty}\frac{\tilde f(\xi'',\eta_{n-1})}{\xi_{n-1}-\eta_{n-1}}\,d\eta_{n-1}.$$

So that $\tilde f = \Pi^+\tilde f + \Pi^-\tilde f$ for all $f(x') \in C_0^\infty(\mathbb{R}^{n-1})$.

Let θ^+ be the operator of the multiplication on $\theta(x_{n-1})$ where $\theta(x_{n-1})=1$ for $x_{n-1} > 0$ and $\theta(x_{n-1}) = 0$ for $x_{n-1} < 0$. Then

$$F(\theta^+ f(x')) = \Pi^+\tilde f,$$

where F is the Fourier transform. It is obvious that Π^+ is a bounded operator in $L_2(\mathbb{R}^{n-1}) \equiv H_0(\mathbb{R}^{n-1})$, because θ^+ is bounded in $L_2(\mathbb{R}^{n-1})$

Lemma 3.1: Operators $\Pi^\pm$ are bounded in $\tilde H'_{\delta,N}(\mathbb{R}^{n-1})$ for $|\delta| < \frac{1}{2}$ and are unbounded for $\delta = \pm\frac{1}{2}$ where $\tilde H'_{\delta,N} = FH'_\delta$ is the image of $H'_\delta(\mathbb{R}^{n-1})$ under the Fourier transform. Here $N \geq 0$ is arbitrary.

The proof of the Lemma 3.1 is given in [6]. The following lemma is a consequence of the Lemma 3.1.

Lemma 3.2: For $|\delta| < \frac{1}{2}$ every function $\tilde f \in \tilde H_{\delta,N}$ can be represented in a unique way in the following form

$$\tilde f = \tilde f_- + \tilde f_+$$

where $\tilde f_- \in \tilde H^-_{\delta,N}$, $\tilde f_+ \in \tilde H^+_{\delta,N}$ and $\tilde f_\pm = \Pi^\pm \tilde f$.

4. Factorization of an elliptic symbol.

Let $b(\xi')$ be a homogeneous function of order α, $b(\xi') \in C^\infty$ for $\xi' \neq 0$. Symbol $b(\xi')$ is called elliptic if $b(\xi') \neq 0$ for $\xi' \neq 0$. By the homogeneous factorization of an elliptic symbol $b(\xi')$ with respect to ξ_{n-1} we mean a representation of $b(\xi')$ in the following form

$$(4.1) \qquad b(\xi') = b_-(\xi')b_+(\xi')$$

where $b_-(\xi')$ and $b_+(\xi')$ satisfy the following conditions :

a) $b_+(\xi')$ $(b_-(\xi'))$ for $\xi' \neq 0$ has an analytic continuation with respect to ξ_{n-1} to the upper (inner) half-plane $\tau = \operatorname{Im} \xi_{n-1} > 0 (\tau = \operatorname{Im} \xi_{n-1} < 0)$.

b) $b_+(\xi'', \xi_{n-1} + i\tau)$ $(b_-(\xi'', \xi_{n-1} + i\tau))$ is continuous in (ξ'', ξ_{n-1}, τ) for $\tau \geq 0$ $(\tau \leq 0)$, $|\xi''| + |\xi_{n-1}| + |\tau| > 0$.

c) $b_+(\xi'', \xi_{n-1} + i\tau)$ $(b_-(\xi'', \xi_{n-1} + i\tau))$ is a homogeneous function in (ξ'', ξ_{n-1}, τ), ord $b_+ = \varkappa$ (ord $b_- = \alpha - \varkappa$), $\varkappa$ is a complex number, in general.

d) $b_+(\xi'', \xi_{n-1} + i\tau) \neq 0$ $(b_-(\xi'', \xi_n + i\tau) \neq 0)$ for $\tau \geq 0$ $(\tau \leq 0)$, $|\xi''| + |\xi_{n-1}| + |\tau| > 0$.

Without loss of generality we can assume that

(4.2) $b_+(0, +1) = 1$.

The order of homogenuity of $b_+(\xi')$ is called the index of the factorization of $b(\xi')$.

Lemma 4.1: Let $b(\xi')$ be an elliptic symbol. Then $b(\xi')$ admit a unique homogeneous factorization (4.1) assuming that the normalization condition (4.2) is fulfilled.

There is the following formula for the index of factorization

$$(4.3) \qquad \varkappa = \frac{\alpha}{2} + \frac{1}{2\pi} \Delta \arg b(\xi'', \xi_{n-1}) \, |\xi'|^{-\alpha} \Big|_{\xi_n = +\infty}^{\xi_n = -\infty} - \frac{1}{2\pi} \ln \left| \frac{b(0,-1)}{b(0,+1)} \right| .$$

The proof of the Lemma 4.1 is given in [6].

Examples of the computation of the index of factorization:

1) Let $b(\xi')$ be a strongly elliptic symbol, i.e $b(\xi') = b_1(\xi') + ib_2(\xi')$, where $b_k(\xi')$ are real, $k=1,2$, and $b_1(\xi') \neq 0$ for $\xi' \neq 0$. Then it follows from (4.3) that

$$\varkappa = \frac{\alpha}{2} + \frac{1}{2\pi i} \ln \frac{b(0,-1)}{b(0,+1)} ,$$

in particular, $\varkappa = \frac{\alpha}{2}$ if $b(0,-1) = b(0,+1)$.

2) Let $b(\xi') = b(-\xi')$ and $n-1 \geq 3$. Then

$$\varkappa = \frac{\alpha}{2} .$$

3) Let $L(\xi')$ be an elliptic polynomial of the order r and $n-1 \geq 3$. Then $r = 2m$ and

$$L(\xi'',\xi_{n-1}) = L_-(\xi'',\xi_{n-1})L_+(\xi'',\xi_{n-1}) \quad ,$$

where $L_\pm(\xi'',\xi_{n-1})$ are polynomials in ξ_{n-1} of the order m.

5. Solution of the mixed problem in the half-space (continuation).

Now we are able to solve the equation (2.7).

Let

$$(5.1) \qquad b_1(\xi')\, b_2^{-1}(\xi') = b_-(\xi')b_+(\xi')$$

be the homogeneous factorization of $b_1(\xi')b_2^{-1}(\xi')$.

Dividing (2.7) by $\hat{b}_-(\xi')$ we obtain

$$(5.2) \qquad \hat{b}_+(\xi')\,\tilde{v}_+(\xi') - \frac{\tilde{v}_-(\xi')}{\hat{b}_-(\xi')} = \hat{b}_+(\xi')\tilde{h}_2(\xi') - \frac{\tilde{h}_1(\xi')}{\hat{b}_-(\xi')} \quad .$$

Let $\varkappa$ be the index of factorization of $b_1b_2^{-1}$, i.e. the order of $b_+(\xi')$ with respect to ξ'.

It follows from the lemma 3.2 that if $|s_0 - \mathrm{Re}\varkappa| < \frac{1}{2}$ then equation (5.2) has a unique solution $\tilde{v}_+ \in \tilde{H}^+_{s_0}(\mathbb{R}^{n-1}_+)$, $\tilde{v}_- \in \tilde{H}^-_{s_0-m_1+m_2}(\mathbb{R}^{n-1}_-)$, where

$$(5.3) \qquad \tilde{v}_+(\xi') = \frac{1}{\overset{*}{b}_+}\,\Pi^+(\hat{b}_+\,\tilde{h}_2 - \frac{\tilde{h}_1}{\hat{b}_-}) \, ,$$

$$(5.4) \qquad \tilde{v}_-(\xi') = -\,\hat{b}_-(\xi')\Pi^-(\hat{b}_+\,\tilde{h}_2 - \tilde{h}_1\,\hat{b}_-^{-1}) \quad .$$

Therefore if $|s-m_2 - \mathrm{Re}\varkappa| < \frac{1}{2}$ then $v(x')$ given by (2.6) is the unique solution of the equation (2.2) in $H'_{s-\frac{1}{2},N}(\mathbb{R}^{n-1})$:

$$(5.5) \qquad \tilde{v}(\xi') = -\,\frac{1}{\hat{b}_2\,\hat{b}_+}\Pi^+(\hat{b}_+\,\tilde{h}_2 - \frac{\tilde{h}_1}{\hat{b}_-}) + \frac{\tilde{h}_1}{\hat{b}_2} \quad .$$

The formula (5.5) can be written in a more symmetric form. We have

$b_1 b_2^{-1} = b_- b_+$. So that $b_1 b_-^{-1} = b_2 b_+$. Further

$$\frac{1}{\hat{b}_2 \hat{b}_+} \prod^+ \frac{\tilde{h}_1}{\hat{b}_-} = \frac{\tilde{h}_1}{\hat{b}_2 \hat{b}_+ \hat{b}_-} - \frac{1}{\hat{b}_2 \hat{b}_+} \prod^- \frac{\tilde{h}_1}{\hat{b}_-} = \frac{\tilde{h}_1}{\hat{b}_1} - \frac{1}{\hat{b}_2 \hat{b}_+} \prod^- \frac{\tilde{h}_1}{\hat{b}_-} .$$

Therefore

$$\tilde{v}(\xi') = - \frac{1}{\hat{b}_2 \hat{b}_+} \prod^+ \hat{b}_+ \tilde{h}_2 - \frac{1}{\hat{b}_1 \hat{b}_-^{-1}} \prod^- \hat{b}_-^{-1} \tilde{h}_1 . \tag{5.6}$$

Thus we have proved the following theorem:

Theorem 5.1: Let the Shapiro-Lopatinskii condition (2.3) be satisfied. Let $\varkappa$ be the index of factorization of $b_1(\xi') b_2^{-1}(\xi')$. Let s and $N \geq 0$ be such that

$$0 < s - m_2 - \operatorname{Re} \varkappa < 1 \tag{5.7}$$

$$S + N \geq \max_{i=1,2} (m_i + \tfrac{1}{2}), \quad s \text{ is non integer if } s \leq \max_{i=1,2} (m_i + \tfrac{1}{2}) .$$

Then there exists a unique solution $u(x) \in H_{s,N}(\mathbb{R}_+^n)$ of the mixed problem (0.8), (0.9) for every $f \in H_{s-2,N}(\mathbb{R}_+^n)$, $g_1 \in H_{s-m_1-\frac{1}{2},N}(\mathbb{R}_+^{n-1})$, $g_2' \in H_{s-m_2-\frac{1}{2},N}(\mathbb{R}_-^{n-1})$.

6. Examples of the mixed problems in the half-space.

6.1) Consider the equation

$$\hat{\Delta} u = f(x) \quad x \in \mathbb{R}_+^n , \tag{6.1}$$

where $\hat{\Delta} = \frac{\partial^2}{\partial x_n^2} + \frac{\partial^2}{\partial x_{n-1}^2} - (|D''| + 1)^2$, $|D''|$ is a pseudo-differential operator with the symbol $|\xi''|$.

Consider the following mixed problem

$$(6.2)\qquad \frac{\partial u}{\partial x_n}\bigg|_{\mathbb{R}^{n-1}_+} = g_1(x'), \quad u\bigg|_{\mathbb{R}^{n-1}_-} = g_2(x') .$$

Here $b_1(\xi') = \sqrt{\xi_{n-1}^2 + |\xi''|^2}$, $b_2(\xi') = 1$.

The factorization of the symbol $b_1(\xi')\, b_2^{-1}(\xi')$ has the following form

$$(6.3)\qquad \sqrt{\xi_{n-1}^2 + |\xi''|^2} = (\xi_{n-1} - i\,|\xi''|)^{\frac{1}{2}}\,(\xi_{n-1} + i\,|\xi''|)^{\frac{1}{2}} ,$$

so that $\varkappa = \frac{1}{2}$. Therefore if

$$(6.4)\qquad \frac{1}{2} < s < \frac{3}{2} \quad \text{and} \quad s + N > \frac{3}{2}$$

then for every $f \in H_{s-2,N}(\mathbb{R}^n_+)$, $g_1 \in H_{s-\frac{3}{2},N}(\mathbb{R}^{n-1}_+)$, $g_2 \in H_{s-\frac{1}{2},N}(\mathbb{R}^{n-1}_-)$

there exists a unique solution $u \in H_{s,N}(\mathbb{R}^n_+)$ of the mixed problem (6.1), (6.2).

6.2) Consider the following mixed problem for the equation (6.1)

$$(6.5)\qquad \hat{\ell}^p\left(\frac{\partial}{\partial x}\right) u\bigg|_{\mathbb{R}^{n-1}_+} = g_1(x'), \quad u\bigg|_{\mathbb{R}^{n-1}_-} = g_2(x') ,$$

where $\ell\left(\frac{\partial}{\partial x}\right) = \sum_{k=1}^{n} a_k \frac{\partial}{\partial x_k}$, a_k are real, $k = 1,\dots,n$, $a_n = 1$ and $p \geq 1$ is an integer.

Here $b_1(\xi') = \ell^p(-i\xi', \sqrt{\xi_{n-1}^2 + |\xi''|^2})$, $b_2(\xi') = 1$ and the index of factorization of $b_1(\xi')\, b_2^{-1}(\xi')$ is equal to

$$\varkappa = p\left(\frac{1}{2} - \frac{1}{\pi} \operatorname{arctg} a_{n-1}\right) .$$

So that the inequalities (5.7) have the following form

$$p\left(\frac{1}{2} - \frac{1}{\pi}\operatorname{arctg} a_{n-1}\right) < s < p\left(\frac{1}{2} - \frac{1}{\pi}\operatorname{arctg} a_{n-1}\right) + 1, \quad s + N > p + \frac{1}{2}$$

We note that if $p = 1$ and $a_{n-1} < 0$ that we can take $N = 0$,

$\frac{3}{2} < s < \frac{3}{2} - \frac{1}{\pi}\operatorname{arctg} a_{n-1}$. But if $a_{n-1} > 0$ then it is necessary to take $N > 0$.

6.3) Consider the following mixed conditions for equation (6.1):

$$\ell_1(\frac{\partial}{\partial x})\, u|_{\mathbb{R}^{n-1}_+} = g_1, \quad \ell_2(\frac{\partial}{\partial x})\, u|_{\mathbb{R}^{n-1}_-} = g_2(x),$$

where $\ell_i(\frac{\partial}{\partial x}) = \sum_{k=1}^{n} a_k^{(i)} \frac{\partial}{\partial x_k}$, $a_n^{(i)} = 1$, $a_k^{(i)}$ are real, $i = 1,2$, $k = 1,\dots,n$.

Then it follows from the formula (4.3) that the index of factorization for this case equal to

$$\varkappa = \frac{1}{\pi} \operatorname{arctg} a_{n-1}^{(2)} - \frac{1}{\pi} \operatorname{arctg} a_{n-1}^{(1)} .$$

So that the inequalities (5.7) gives

$$s + N > \frac{3}{2}, \quad \varkappa < s < \varkappa + 1 .$$

7. Asymptotic behaviour of the solution of the mixed problem near the hyperplane $x_n = x_{n-1} = 0$.

The formulas (2.2) and (5.6) give an explicit representation for the unique solution of the problem (0.8), (0.9). Suppose that $f, g_1 g_2$ are C^∞ functions equal to zero in the neighbourhood of the infinity.
Then taking an expansion of $b_1(\xi')$, $b_2(\xi')$, $b_+(\xi')$ for $\xi_{n-1} \to +\infty$ and for $\xi_{n-1} \to -\infty$ we can obtain the following expansion for $u(x',x_n)$ for $x_{n-1}^2 + x_n^2 \to 0$ (see [6] for the details):

$$u(x',x_n) = b_2^{-1}(0,+1) d_0(x'')\, z_1^{\varkappa+m_2} - e^{-\varkappa\pi i}\, b_2^{-1}(0,-1) d_0(x'')\, z_2^{\varkappa-m_2} +$$

(7.1)

$$+ \sum_{i=1}^{2} \sum_{r=0}^{M} \sum_{p+k=1}^{M} d_{ipkr}\, (x'') x_n^p\, z_i^{\varkappa+m_2+k} \ln^r z_i + u_M(x',x_n),$$

where M is arbitrary, $u_M \in C^{M+m_2+[\operatorname{Re}\varkappa]}(\overline{\mathbb{R}^n_+})$,

$z_1 = -(x_{n-1} + x_n\, \lambda(0,+1))$, $z_2 = x_{n-1} - x_n\, \lambda(0,-1)$,

$$d_0(x'') = e^{-i\frac{\pi}{2}(\varkappa+m_2)}\,\Gamma(-\varkappa-m_2)\cdot$$

$$\int_{-\infty}^{\infty}(\hat{b}_-^{-1}(\xi')\tilde{h}_1(\xi') - \hat{b}_+(\xi')h_2(\xi'))e^{-i(x'',\xi'')}\,d\xi''d\xi_{n-1},\quad d_{ipkr}(x'') \in C^\infty(\mathbb{R}^{n-2})$$

and also can be calculated explicitly. Above we have supposed that $\varkappa$ is noninteger. If $\varkappa$ is an integer and $\varkappa + m_2 \geq 0$ then it is necessary to replace $z_i^{\varkappa+m_2}$ by $z_i^{\varkappa+m_2}\ln z_i$ in the two first terms in (7.1).

Therefore it follows from (7.1) that $u(x',x_n)$ has in the neighbourhood of the hyperplane $x_n = x_{n-1} = 0$ the same smoothness as $r^{\varkappa+m_2}$ or $r^{\varkappa+m_2}\ln r$ if $\varkappa + m_2 \geq 0$ and $\varkappa$ is an integer, $r=\sqrt{x_{n-1}^2+x_n^2}$.

8. Mixed problem for second order elliptic equation in a bounded domain.

Consider now a mixed problem (0.1),(0.2),(0.3). Let $B_k(x, D)$ satisfies on $\bar{\Gamma}_k$ (k=1,2) the Shapiro-Lopatinskii condition in a corresponding local system of coordinates. Let $x_0'' \in \Gamma_0$ be arbitrary and let $\varkappa(x_0'')$ be the index of the factorization (4.1) in a local system of coordinates. It can be proven that $\varkappa(x_0'')$ does not depend on a choice of a system of coordinates and that $\varkappa(x'')$ is a smooth function on Γ_0.

Denote by $H_{s,N}(G)$, $H'_{s,N}(\Gamma_k)$, $k = 1,2$ Sobolev's spaces with weights corresponding to $H_{s,N}(\mathbb{R}_+^n)$, $H'_{s,N}(\mathbb{R}_+^{n-1})$, $H'_{s,N}(\mathbb{R}_-^{n-1})$.

Suppose that

$$(8.1)\qquad \max_{x''\in\Gamma_0} \operatorname{Re}\varkappa(x'') - \min_{x''\in\Gamma_0} \operatorname{Re}\varkappa(x'') < 1 \quad .$$

Then on the base of the Theorem 5.1 the following theorem can be proved by the usual technique of "frozen" coefficients.

Theorem 8.1 : Let $B_k(x, D)$ satisfy the Shapiro-Lopatinskii condition on $\bar{\Gamma}_k$(k=1,2). Let s and N satisfy the following inequalities:

$$(8.2)\quad 0 < s - \operatorname{Re}\varkappa(x'') - m_2 < 1,\quad \forall x'' \in \Gamma_0,\quad s + N > \max_{k=1,2}(m_k + \tfrac{1}{2}) \quad \text{and}$$

s is noninteger if $s \leq \max_{k=1,2}(m_k + \frac{1}{2})$.

Then mixed problem (0.1), (0.2), (0.3) is normally solvable for $u \in H_{s,N}(G)$, $g_k \in H_{s-m_k-\frac{1}{2},N}(\Gamma_k)$, $k = 1,2$, $f \in H_{s-2,N}(G)$.

In general $\mathscr{X}(x')$ is an arbitrary function on Γ_0 so that the condition (8.1) can be not fulfilled. Then we shall introduce a space of piecely constant order of smoothness in a following way:

Let $\{\varphi_j(x)\ u_j\}$ be the partition of unity in $\bar{G}$ and $\{s_j\}$ be real numbers such that $|s_i - s_j| < 1$ if $\bar{U}_i \cap \bar{U}_j \neq \phi$. Then we denote by $H_{\{s_j\},N}(G)$ a functional space with the following norm

$$(8.3) \quad \|u\|_{(s_j),N} = \sum_j \|\varphi_j u\|_{s_j,N} ,$$

where $\|\varphi_j u\|_{s_j,N}$ is the norm in the space $H_{s_j,N}(G)$.

In the same way we introduce spaces $H'_{\{s_j\},N}(\Gamma_k)$, $k = 1,2$, with the norms $\|v\|'_{(s_j),N} = \sum_j \|p'\varphi_j v\|'_{s_j,N}$ where $p'\varphi_j$ is the restriction of φ_j to $\bar{\Gamma}_k$.

It is clear that $p'\varphi_j = 0$ if $\bar{U}_j \cap \bar{\Gamma}_k = \phi$.

Now let $0 < s_j - \mathrm{Re}\mathscr{X}(x'') - m_2 < 1$ for $x'' \in \bar{U}_j \cap \Gamma_0$, $s_j + N > \max_{k=1,2} (m_k + \frac{1}{2})$ and s_j is noninteger if $s_j \leq \max_{k=1,2} (m_k + \frac{1}{2})$. Then the problem (0.1), (0.2), (0.3) is normally solvable for $u \in H_{\{s_j\},N}(G)$, $f \in H_{\{s_j\}-2,N}(G)$, $g_k \in H'_{\{s_j\}-m_k-\frac{1}{2}}(\Gamma_k)$, $k = 1,2$.

9. Generalized mixed problem.

It was shown in the section 2 that the solution of the mixed problem (0.8), (0.9) can be reduced to the solution of equation (2.7). Set $t = s - m_2 - \frac{1}{2}$. If $t - \mathrm{Re}\mathscr{X} = \delta, |\delta| < \frac{1}{2}$ then equation (2.7) has a unique solution (5.3), (5.4) (see the section 5). Now we shall study equation (2.7) for arbitrary t.

Let

(9.1) $t - \mathrm{Re}\mathscr{X} = -m + \delta$, $m > 0$ is an integer, $|\delta| < \frac{1}{2}$. Consider the equation (2.7) for $\tilde{h}(\xi') = \frac{1}{\hat{b}_-} (\hat{b}_1 \hat{b}_2^{-1} \tilde{h}_2 - \tilde{h}_1) \in \tilde{H}'_{-m+\delta,N}(\mathbb{R}^{n-1})$

and we look for $\hat{b}_+\tilde{v}_+ \in \tilde{H}^+_{-m+\delta,N}$, $\frac{\tilde{v}_-}{\hat{b}_-} \in \tilde{H}^-_{-m+\delta,N}$.

Let $Q(\xi'',\xi_{n-1}) = \sum_{k=0}^{m} q_k(\xi'')\, \xi_{n-1}^k$ be a homogeneous function of the order m, C^∞ for $\xi'' \neq 0$ which is a polynomial with respect to ξ_{n-1}.

Let $Q(\xi'',\xi_{n-1}) \neq 0$ for $\xi' = (\xi'',\xi_{n-1}) \neq 0$. Then $\frac{\tilde{h}}{\tilde{Q}} \in \tilde{H}'_{\delta,N}(\mathbb{R}^{n-1})$, so that $\frac{\tilde{h}}{\tilde{Q}} = \Pi^+ \frac{\tilde{h}}{\tilde{Q}} + \Pi^- \frac{\tilde{h}}{\tilde{Q}}$ as it follows from Lemma 3.2. Therefore the equation (2.7) can be rewritten in the following form

$$(9.2)\qquad \hat{b}_+\tilde{v}_+ - \hat{Q}\,\Pi^+ \frac{\tilde{h}}{\tilde{Q}} = \hat{b}_-^{-1}\tilde{v}_- + \hat{Q}\,\Pi^- \frac{\tilde{h}}{\tilde{Q}} \in \tilde{H}^+_{-m+\delta} \cap \tilde{H}^-_{-m+\delta} .$$

So that the Liouville theorem gives that

$$(9.3)\qquad \hat{b}_+\tilde{v}_+ - \hat{Q}\,\Pi^+ \frac{\tilde{h}}{\tilde{Q}} = \sum_{k=1}^{m} \tilde{c}_k(\xi'')\, \xi_n^{k-1} ,$$

where $\tilde{c}_k(\xi'') \in \tilde{H}''_{t_k}(\mathbb{R}^{n-2})$, $t_k = t - \operatorname{Re}\varkappa + k - \frac{1}{2}$, $1 \leq k \leq m$.

Therefore in the case (9.1) the equation (2.7) has solution but it is not unique and depend on m arbitrary functions $\tilde{c}_k(\xi') \in \tilde{H}''_{t_k}(\mathbb{R}^{n-2})$.

Consider now the case

(9.4) $t - \operatorname{Re}\varkappa = m + \delta$, $m > 0$ is an integer and $|\delta| < \frac{1}{2}$. Since $\tilde{h}(\xi') \in \tilde{H}'_{m+\delta,N}(\mathbb{R}^{n-1}) \subset \tilde{H}'_{\delta,N}(\mathbb{R}^{n-1})$ we have as in the section 5 that if the solutions $\tilde{v}_+(\xi') \in \tilde{H}^+_{m+\delta,N}$, $\tilde{v}_-(\xi') \in \tilde{H}^-_{m+\delta,N}$ exist then they must be given by the formulas (5.4). But, in general, $\Pi^\pm\tilde{h}$ does not belong to $\tilde{H}^\pm_{m+\delta,N}$, $m > 0$. There is a following simple formula for the Cauchy integral:

$$(9.5)\qquad \Pi^+\tilde{h} = \sum_{k=1}^{m} \frac{1}{\hat{\Lambda}_+^k} \frac{i}{2\pi} \int_{-\infty}^{\infty} \hat{\Lambda}_+^{k-1}(\xi'',\xi_{n-1})\tilde{h}(\xi'',\xi_{n-1})d\xi_{n-1} + \frac{1}{\hat{\Lambda}_+^m}\Pi^+\hat{\Lambda}_+^m\tilde{h} ,$$

where

$$(9.6)\qquad \Lambda_+(\xi') = \xi_{n-1} + i\,|\xi''| .$$

It is obvious that $\frac{1}{\hat{\Lambda}_+^m}\Pi^+ \hat{\Lambda}_+^m \tilde{h} \in \tilde{H}^+_{m+\delta,N}$, but the sum in (9.5) belongs to $\tilde{H}^+_{m+\delta,N}$ only if for almost all ξ'' we have

(9.7) $\int_{-\infty}^{\infty} \hat{\Lambda}_{+}^{k-1} (\xi'',\xi_{n-1}) \tilde{h}(\xi'',\xi_{n-1}) d\xi_{n-1} = 0 \ , \ 1 \leq k \leq m.$

Therefore in the case (9.4) there is the uniqueness of the solution of equation (2.7) but for the existence it is necessary to satisfy the condition (9.7).

We note that if $t - \mathrm{Re}\,\varkappa = \frac{1}{2} \pmod{k}$, k is integer then the image of the operator corresponding to the equation (2.7) is not close.

Therefore only if $|t - \mathrm{Re}\,\varkappa| < \frac{1}{2}$, $t = s - m_2 - \frac{1}{2}$, then the mixed problem (0.8), (0.9) has a unique solution. If $t - \mathrm{Re}\,x = -m + \delta$, $m > 0$, $|\delta| < \frac{1}{2}$ then the mixed problem has an infinitely dimensional kernel. If $t - \mathrm{Re}\,\varkappa = m + \delta$, $m > 0$, $|\delta| < \frac{1}{2}$ then the mixed problem (0.8), (0.9) has an infinitely dimensional cokernel.

Therefore for $|t - \mathrm{Re}\,x| > \frac{1}{2}$ we shall introduce some generalization of the mixed problem (0.8), (0.9) in order to obtain a normally solvable problem. Consider at first the case (9.1). Then in order to get rid of the kernel of equation (2.7) we shall impose additional boundary conditions on $\mathbb{R}^{n-2}$. Thus we shall consider the following problem

(9.9) $\hat{L}(D)u = f, \quad x \in \mathbb{R}^n_+ ,$

(9.10) $p'_+ \hat{B}_1(D)u = g_1, \quad p'_- \hat{B}_2(D)u = g_2 ,$

(9.11) $p'' B_k^- \ell u = g_{k0}(x''), \ 1 \leq k \leq m ,$

where $B_k^-(D)$ are pseudodifferential operators and p'' is the restriction operator on $\mathbb{R}^{n-2}$. We suppose that the symbols $B_k^-(\xi',\xi_n)$ are analytic in ξ_n for $\mathrm{Im}\,\xi_n < 0$ and $B_k^-(\xi',\xi_n) = \beta_k < s - 1$, so that the restriction $p''\hat{B}_k^- \ell u$ exists and does not depend on a choice of ℓu.

It may be shown that we can choose $B_k^-(\xi',\xi_n)$ such that the problem (9.9), (9.10), (9.11) has a unique solution $u \in H_{s,N}(\tilde{\mathbb{R}}_+)$, $s = m_2 + \frac{1}{2} - \mathrm{Re}\,\varkappa - m + \delta$

for arbitrary $f \in H_{s-2,N}(\mathbb{R}^n_+)$, $g_k \in H'_{s-m_k-\frac{1}{2},N}(\mathbb{R}^{n-1}_\pm)$, $k = 1,2,$
$g_{j0} \in H''_{s-\beta_j-1}(\mathbb{R}^{n-2})$, $1 \leq j \leq m$.

Consider now the case (9.4). In this case we shall add to the boundary conditions (0.9) m operators of the type

$$\hat{G}_k(D')(v_k(x'') \times \delta(x_{n-1})) = \frac{1}{(2\pi)^{n-1}} \int_{-\infty}^{\infty} \hat{G}_k(\xi'',\xi_{n-1}) \tilde{v}_k(\xi'') e^{-i(x',\xi')} d\xi'' d\xi_{n-1} ,$$

$$1 \leq k \leq m,$$

in order to compensate for the cokernel of the problem (0.8), (0.9). Thus we shall consider the following problem

$$(9.12) \quad L(D)u = f(x),\ x \in \mathbb{R}^n_+ ,$$

$$(9.13) \quad p'_+(B_1(D)u + \sum_{k=1}^{m} G_{k1}(v_k(x'') \times \delta(x_{n-1}))) = g_1(x') ,$$

$$(9.14) \quad p'_-(B_2(D)u + \sum_{k=1}^{m} G_{k2}(v_k(x') \times \delta(x_{n-1})) = g_2(x') ,$$

where $u \in H_{s,N}(\mathbb{R}^n_+)$, $v_k(x') \in H''_{s_k}(\mathbb{R}^{n-2})$ are unknown, $\mathrm{ord}_\xi G_{ki}(\xi'',\xi_{n-1}) = m_i + \gamma_k < -1$, $s_k = s + \gamma_k - \frac{1}{2}$.

We can choose $G_{ki}(\xi')$ such that this problem has a unique solution.

We note that for second order elliptic equations the generalized mixed problems as (9.9), (9.10), (9.11) and (9.12), (9.13), (9.14) are not natural because we can choose the functional space $H_{s,N}(\mathbb{R}^n_+)$ in such a way that the usual mixed problem (0.8), (0.9) has a unique solution. But in the case of mixed problem for higher order elliptic equations and systems of elliptic equations it is impossible in general to choose such functional space. So that in this case we are imposed to consider generalized mixed problem. In general, for higher order elliptic equations and systems of equations we need at the same time both additional boundary conditions as in (9.11) and additional potentials as in (9.13), (9.14).

10. The Riemann-Hilbert problem for the system of functions

As in the case of second order elliptic equations the mixed problem for system of elliptic equations or for a single higher order elliptic equation is connected with the solution of the Riemann-Hilbert problem

(10.1) $\quad \hat{A}(\xi'',\xi_{n-1})\,\tilde{v}_+(\xi') - \tilde{v}_-(\xi') = \tilde{h}(\xi')$

where $A(\xi') = \|a_{ij}(\xi')\|_{i,j=1}^{p}$ is an elliptic homogeneous of order α matrix, C^∞ for $\xi' \neq 0$ and $v_+ \in H^+_{t,N}$, $v_- \in H^-_{t-\alpha,N}$.

Let $\omega = \frac{\xi''}{|\xi''|}$ be fixed. Consider the following Riemann-Hilbert problem in one variable

(10.2) $A(\omega,\xi_{n-1})\,\tilde{v}_+(\xi_{n-1}) - \tilde{v}_-(\xi_{n-1}) = \tilde{h}(\xi_{n-1})$,

where $\tilde{v}_+ \in \tilde{H}^+_{t,N}(\mathbb{R}^1_+)$, $\tilde{v}_- \in \tilde{H}^-_{t-\alpha,N}(\mathbb{R}^1_-)$, $\tilde{h} \in \tilde{H}_{t-\alpha,N}(\mathbb{R}^1)$.

We shall show later that the study of the Riemann-Hilbert problem (10.1) is based on an investigation of the family of Riemann-Hilbert problem (10.2) depending on the parameter $\omega = \frac{\xi''}{|\xi''|}$.

The Riemann-Hilbert (10.2) can be written as a system of pseudodifferential operators on the half-line:

(10.3) $\quad p_+'\, A(\omega,D_{n-1})v_+(x_{n-1}) = p_+' h$,

where we have taken the inverse Fourier transform for (10.2) and then the restriction on the half-line $\mathbb{R}^1_+$; We note that $p_+' v_- = 0$, $p_+' h \in H_{t-\alpha,N}(\mathbb{R}^1_+)$.

Let $\Lambda_\pm(\xi_{n-1}) = \xi_{n-1} \pm i$. We put $\tilde{u}_+ = \Lambda_+^t(\xi_{n-1})\,\tilde{v}_+$, $\tilde{u}_- = \Lambda_-^{(t-\alpha)}\tilde{v}_-$, $\tilde{h}_1 = \Lambda_-^{t-\alpha}\tilde{h}$. Then we obtain the following system which is equivalent to (10.2):

(10.4) $\quad A_0(\omega,\xi_{n-1})\,\tilde{u}_+(\xi_{n-1}) - \tilde{u}_-(\xi_{n-1}) = \tilde{h}_1(\xi_{n-1})$,

where $A_0(\omega,\xi_{n-1}) = A(\omega,\xi_{n-1})\,\Lambda_+^{-t}\,\Lambda_-^{t-\alpha}$, $\tilde{u}_+ \in \tilde{H}^+_{0,N}$, $\tilde{u}_- \in \tilde{H}^-_{0,N}$, $\tilde{h}_1 \in \tilde{H}_{0,N}(\mathbb{R}^1)$

The solution of the system (10.4) is equivalent to the solution of the system

(10.5) $\Pi^+ A_0(\omega,\xi_{n-1})\, \tilde{u}_+ = \tilde{h}_+$,

where $\tilde{h}_+ = \Pi^+ \tilde{h}_1$.

If we take the inverse Fourier transform we shall obtain from (10.5):

(10.6) $\theta^+ A_0(\omega,D_{n-1})\, u_+(x_{n-1}) = h_+(x_{n-1})$,

where $A_0(\omega,D_{n-1})\, u_+ = \frac{1}{2\pi} \int_{-\infty}^{\infty} A_0(\omega,\xi_{n-1})\, \tilde{u}_+(\xi_{n-1})\, e^{-ix_{n-1}\xi_{n-1}}\, d\xi_{n-1}$, θ^+ is the operator of multiplication on $\theta(x_n)$, $u_+ \in H^+_{0,N}$, $h_+ \in H^+_{0,N}$.

The first problem which we shall consider will be to find the condition for $\theta^+ A_0$ to be a Fredholm operator in $H^+_{0,N}(\mathbb{R}^1)$.

Let $a_\pm = \lim_{\xi_{n-1}\to\pm\infty} A_0(\omega,\xi_{n-1})$ and let $a_1,\dots,a_p$ be the eigenvalues of the matrix $a_+^{-1}a_-$ with their multiplicity taken into account.

<u>Theorem 10.1</u>: <u>The necessary and sufficient condition for $\Pi^+ A_0(\omega,\xi_{n-1})$ to be a Fredholm operator in $\tilde{H}^+_{0,N}$ is the following:</u>

(10.7) <u>There are no real negative numbers among the eigenvalues</u> a_j, $1 \le j \le p$.

<u>Proof</u>: (see [5], [6]): Set

(10.8) $b = \frac{1}{2\pi i} \ln(a_+^{-1}a_-)$, $b_\pm(\xi_{n-1}) = \exp\{\ln(\xi_{n-1} \pm i)b\}$.

Let b_j, $1 \le j \le p$, be the eigenvalues of the matrix b. By using the condition (10.7) we can choose the branch of the logarithm such that

(10.9) $-\frac{1}{2} < \mathrm{Re}\, b_j < \frac{1}{2}$, $1 \le j \le p$.

It is easy to see that the matrix $a_+ b_-^{-1}(\xi_{n-1}) b_+(\xi_{n-1})$ has the same limits as the matrix $A_0(\omega,\xi_{n-1})$ when $\xi_{n-1} \to \pm\infty$.

Set

(10.10) $$P_{\pm}\tilde{u}_+ = b_{\pm}^{-1}\Pi^+ b_{\pm}\tilde{u}_+$$

Lemma 10.2: (see [5], [6]) If conditions (10.9) holds then the operators $P_{\pm}$ are bounded in $H_{0,N}(\mathbb{R}^1)$.

Now using the Lemma 10.2 and the following relation

$$P_+A_0^{-1}a_+ = A_0^{-1}a_+P_- + C_1 ,$$

where C_1 is a Hilbert-Schmidt operator, we can prove that the operator $\tilde{\tilde{R}}g_+ = P_+A_0^{-1}a_+P_-a_+^{-1}\tilde{g}_+$ is a left and right regularizer of Π^+A_0 (see [5],[6] for the details):

$$\Pi^+A_0\tilde{\tilde{R}}g_+ = \tilde{g}_+ + T_1\tilde{g}_+ ,$$

(10.11)

$$\tilde{R}\Pi^+A_0\tilde{g}_+ = \tilde{g}_+ + T_2\tilde{g}_+,$$

where T_1,T_2 are compact operators in $\tilde{H}_{0,N}$. It follows from (10.11) that Π^+A_0 is a Fredholm operator in $\tilde{H}^+_{0,N}$.

It is easy to rewrite the condition (10.7) for the case of the equation (10.3) or (10.2). Set $A^{(0)}(\xi'',\xi_{n-1}) = A(\xi'',\xi_{n-1})(\xi_n - i|\xi''|)^{-\alpha}$.

Let $a_j^{(0)}$, $1 \le j \le p$, be the eigenvalues of the matrix $A^{(0)}(0,-1)[A^{(0)}(0,+1)]^{-1}$.

Then the necessary and sufficient conditions for the operator (10.2) to be a Fredholm operator from $\overset{+}{H}_{t,N}$ to $H_{t-\alpha,N}(\mathbb{R}^1_+)$ are the following:

(10.12) $$t - \mathrm{Re}\,\frac{1}{2\pi i}\ln a_j^{(0)} \neq \frac{1}{2} \pmod k ,$$

where k is an arbitrary integer.

11. Normal solvability of pseudodifferential operator of the half-line (continuation)

In this section we give two others methods to prove the normal solvability of the equation (10.6).

11.1) Set $A_1(\omega,\xi_{n-1}) = A_0(\omega,\xi_{n-1}) - a_+ b_-^{-1}(\xi_{n-1}) b_+(\xi_{n-1})$ and let $\chi(x_{n-1}) \in C_0^\infty(\mathbb{R}^1)$, $\chi(x_{n-1}) = 1$ for $|x_{n-1}| \leq 1$. We note that $(1-\chi(x_{n-1}))\theta(x_{n-1}) \in C^\infty(\mathbb{R}^1)$ and that the operators $\chi(x_{n-1})A_1(\omega,D_{n-1})$, $A_1(\omega,D_{n-1})\chi(x_{n-1})$ and the commutators $[\chi(x_{n-1}), A_0(\omega,D_{n-1})]$, $[\chi(x_{n-1}), a_+ b_-^{-1}(D_{n-1}) b_+(D_{n-1})]$ are Hilbert-Schmidt operators.

Denote by R_0 the following operator

$$(11.1)\quad R_0 g_+ = \chi(x_{n-1})F^{-1}_{\xi_{n-1}} b_+^{-1}(\xi_{n-1}) \Pi^+ b_- a_+ \tilde{g}_+ + (1-\chi(x_{n-1}))\cdot$$

$$\theta^+ F^{-1}_{\xi_{n-1}} A_0^{-1}(\omega,\xi_{n-1})\tilde{g}_+ \ .$$

Taking into account that the operator $b_+^{-1}(\xi_{n-1})\Pi^+ b_- a_+^{-1}$ is the inverse of $\Pi^+ a_+ b_-^{-1} b_+$ we have

$$(11.2)\quad \theta^+ A_0 R_0 g_+ = g_+ + T_1 g_+ \ ,$$

$$R_0\theta^+ A_0 g_+ = g_+ + T_2 g_+ \ ,$$

where T_1, T_2 are compact operators in $\overset{+}{H}_{0,N}(\mathbb{R}^1_+)$.

11.2) The previous two methods of constructing regularizer for the equation (10.6) show that if $a_+ = a_-$ then the regularizer of $\theta^+A_0(\omega,D_{n-1})$ has the form $\theta^+A_0^{-1}(\omega,D_{n-1})$. If $a_+ \neq a_-$ then the regularizer of $\theta^+A_0(\omega,D_{n-1})$ is not an operator of the same class. In this section we extend the class of pseudodifferential operators of order zero on the half-line in such a way that the extended class contains the regularizers of the operators $\theta^+A_0(\omega,D_{n-1})$.

Let $M(t)$ be a matrix which belongs to C^∞ for $t > 0$ and satisfies the estimates

$$(11.3)\quad t^k\left|\frac{d^kM(t)}{dt^k}\right| \le \begin{cases} c_\varepsilon\, t^{-\delta-\varepsilon} & \text{for } 0 < t < 1,\ 0 \le k < \infty, \\ c_\varepsilon\, t^{-1+\delta+\varepsilon} & \text{for } 1 \le t < \infty,\ 0 \le k < \infty. \end{cases}$$

In (11.3), δ is fixed, $0 < \delta < \frac{1}{2}$ and $\varepsilon > 0$ is arbitrary. Denote by $\hat{M}(z)$ the Mellin transform of $M(t)$:

$$(11.4)\quad \hat{M}(z) = \int_0^\infty M(t)\, t^{z-1} dt.$$

Then $\hat{M}(z)$ is analytic in the strip $\delta < \operatorname{Re} z < 1-\delta$ and it satisfies the estimates

$$(11.5)\quad |z^m\hat{M}(z)| \le C_{\varepsilon m},\ \delta+\varepsilon \le \operatorname{Re} z \le 1-\delta-\varepsilon,\ \forall m \ge 0,\ \forall \varepsilon > 0$$

Conversely if $\hat{M}(z)$ is analytic in the strip $\delta < \operatorname{Re} z < 1-\delta$ and satisfies (11.5), then $M(t) \in C^\infty$ for $t > 0$, and the estimates (11.3) holds.

Let M be the integral operator defined by

$$(11.6)\quad Mu_+ = \int_0^\infty M\left(\frac{t}{\tau}\right)\frac{u_+(\tau)}{\tau}\, d\tau,\quad \forall u \in C_0^\infty(\mathbb{R}_+^1).$$

Then $\widehat{Mu_+} = \hat{M}(z)\cdot\hat{u}_+(z)$,

where $\hat{u}_+(z)$ is the Mellin transform of $u_+(x_{n-1})$. We shall call M the Mellin operator and $\hat{M}(z)$ will be called the symbol of the operator M. The class of the operators (11.6), where $M(t)$ satisfies the estimates (11.3), will be denoted by $\mathfrak{M}$.

We shall consider the following class of operators on the half-line $\mathbb{R}_+^1$:

$$(11.7)\quad \theta^+A_0(\omega, D_{n-1})\, u_+(x_{n-1}) + \theta^+\chi(x_{n-1})\, M u_+ + \theta^+ Tu_+,$$

where $A_0(\omega, D_{n-1})$ is the pseudodifferential operator (ψd.0) on $\mathbb{R}^1$ with the symbol $A_0(\omega, \xi_{n-1})$, M is an integral operator of the class $\mathfrak{M}$ and T is a compact operator in $H_{0,N}(\mathbb{R}^1)$.

Since $A(\xi'',\xi_{n-1})$ is a homogeneous matrix of class C^∞ for $\xi' \neq 0$ the limit $a_\pm = \lim_{\xi_{n-1}\to\pm\infty} A_0(\omega,\xi_{n-1})$ always exists. By the symbol of the operator (11.7) we mean the pair of matrices.

(11.8) $A_0(\omega,\xi_n)$ on the closed line $-\infty \leq \xi_n \leq +\infty$

and

(11.9) $a^+ \dfrac{1}{1-e^{2\pi iz}} - a_- \dfrac{e^{2\pi i z}}{1-e^{2\pi iz}} + \hat{M}(z)$ on the closed line $\mathrm{Re} z = \frac{1}{2}$,

where $a_\pm = \lim_{\xi_{n-1}\to\pm\infty} A_0(\omega,\xi_n)$.

<u>Theorem 11.1:</u> (see [5], [6]) <u>Let</u> $\mathfrak{A}_1 = \theta^+ A_1(\omega,D_{n-1}) + \theta^+ \chi(x_{n-1}) M_1 + \theta^+ T_1$ <u>and</u> $\mathfrak{A}_2 = \theta^+ A_2(\omega,D_{n-1}) + \theta^+ \chi(x_{n-1}) M_2 + \theta^+ T_2$ <u>be two operators of the form (11.7) with the symbols</u>

$$A_1(\omega,\xi_{n-1}), \quad a_1^+ \frac{1}{1-e^{2\pi iz}} - a_1^- \frac{e^{2\pi iz}}{1-e^{2\pi iz}} + \hat{M}_1(z)$$

<u>and</u>

$$A_2(\omega,\xi_{n-1}), \quad a_2^+ \frac{1}{1-e^{2\pi iz}} - a_2^- \frac{e^{2\pi iz}}{1-e^{2\pi iz}} + \hat{M}_2(z),$$

where $a_1^\pm = \lim_{\xi_{n-1}\to\pm\infty} A_1(\omega,\xi_{n-1})$, $a_2^\pm = \lim_{\xi_{n-1}\to\pm\infty} A_2(\omega,\xi_{n-1})$.

<u>Then their product</u> $\mathfrak{A}_3$ <u>can be put in the form (11.7):</u>

$$(11.10) \quad \mathfrak{A}_3 = \mathfrak{A}_1\mathfrak{A}_2 = \theta^+ A_3(\omega,D_{n-1}) + \theta^+ \chi(x_{n-1}) M_3 + \theta^+ T_3$$

<u>and</u> $\mathfrak{A}_3$ <u>has the symbol:</u>

$$A_3(\omega,\xi_{n-1}) \quad , \quad a_3^+ \frac{1}{1-e^{2\pi iz}} - a_3^- \frac{e^{2\pi iz}}{1-e^{2\pi iz}} + \hat{M}_3(z),$$

<u>which is equal to the product of the symbols of</u> $\mathfrak{A}_1$ <u>and</u> $\mathfrak{A}_2$; <u>that is</u>

(11.11) $$A_3(\omega,\xi_{n-1}) = A_1(\omega,\xi_{n-1})\, A_2(\omega,\xi_{n-1}) ,$$

$$a_3^+ \frac{1}{1-e^{2\pi iz}} - a_3^- \frac{e^{2\pi iz}}{1-e^{2\pi iz}} + \hat{M}_3(z) = (a_1^+ \frac{1}{1-e^{2\pi iz}} - a_1^- \frac{e^{2\pi iz}}{1-e^{2\pi iz}} + \hat{M}_1(z))$$

$$\cdot (a_2^+ \frac{1}{1-e^{2\pi iz}} - a_2^- \frac{e^{2\pi iz}}{1-e^{2\pi iz}} + \hat{M}_2(z)) ,$$

where $a_3^\pm = \lim_{\xi_{n-1}\to\pm\infty} A_3(\omega,\xi_{n-1}) = a_1^\pm a_2^\pm$.

As a consequence of the Theorem 11.1 we obtain the following condition for the normal solvability of an operator $\mathcal{A}$ of the form (11.7):

For $\mathcal{A}$ to be a Fredholm operator it is necessary and sufficient that the determinant of its symbol (11.8), (11.9) is now zero:

(11.12) $$\det A_0(\omega,\xi_{n-1}) \neq 0, \quad -\infty \leq \xi_{n-1} \leq +\infty , \quad \det a_\pm \neq 0,$$

(11.13) $$\det \left(\frac{a_+}{1-e^{2\pi iz}} - \frac{a_- e^{2\pi iz}}{1-e^{2\pi iz}} + \hat{M}(z)\right) \neq 0 , \ \mathrm{Re}\, z = \frac{1}{2} .$$

If conditions (11.12), (11.13) are fulfilled we can construct a regularizer R of the operator $\mathcal{A}$ in the following form:

(11.14) $$R\, u_+ = \theta^+ A_0^{-1}(\omega,D_{n-1}) u_+ + \theta^+ \chi(x_n)\, M_1 u_+ ,$$

where

(11.15) $$M_1(z) = \left(\frac{a_+}{1-e^{2\pi iz}} - \frac{a_- e^{2\pi iz}}{1-e^{2\pi iz}} + M(z)\right)^{-1} - \left(\frac{a_+^{-1}}{1-e^{2\pi iz}} - \frac{a_-^{-1} e^{2\pi iz}}{1-e^{2\pi iz}}\right) .$$

In particular, let $\mathcal{A} = \theta^+ A_0(\omega,D_{n-1})$, that is, $M = 0$, $T = 0$. Then $\theta^+ A_0(\omega,D_{n-1})$ is a Fredholm operator if

(11.16) $$\det A_0(\omega,\xi_{n-1}) \neq 0, \quad -\infty \leq \xi_n \leq \infty , \quad \det a_\pm \neq 0,$$

$$(11.17)\qquad \det\left(\frac{a_+}{1-e^{2\pi iz}} - \frac{a_- e^{2\pi iz}}{1-e^{2\pi iz}}\right) \neq 0, \ \mathrm{Re}z = \frac{1}{2} .$$

Since $z = \frac{1}{2} + i\tau$, we have $e^{2\pi iz} = - e^{-2\pi\tau}$.

Condition (11.17) is equivalent to the following

$$(11.18)\qquad \det(e^{2\pi\tau}\, I + a_+^{-1}\, a_-) \neq 0 \quad \text{for} \quad -\infty < \tau < +\infty .$$

Hence the conditions (10.7) and (11.17) are equivalent.

Let $t = \frac{1}{1+e^{-2\pi\tau}}$. Then $0 \leq t \leq 1$ when $-\infty \leq \tau \leq +\infty$.

The condition (11.17) can be written in the following form (c.f. [13])

$$(11.19)\qquad \det(ta_+ + (1-t)a_-) \neq 0 \quad \text{for} \quad 0 \leq t \leq 1 .$$

We note that $ta_+ + (1-t)a_-$ is the segment in the space of matrices which connects $a_+ = A_0(\omega,+\infty)$ with $a_- = A_0(\omega,-\infty)$.

Now we shall find the index of a Fredholm operator $\mathcal{A}$ of the form (11.7).

Theorem 11.2: *When (11.12), (11.13) hold then the index of the operator $\mathcal{A}$ is given by*

$$(11.20)\qquad \begin{aligned} \operatorname{ind}\mathcal{A} = {} & \frac{1}{2\pi}\,\Delta\arg\det\, A_0(\omega,\xi_{n-1})\Big|_{\xi_{n-1}=+\infty}^{\xi_{n-1}=-\infty} + \\ & + \frac{1}{2\pi}\,\Delta\arg\det\left(\frac{a_+}{1-e^{2\pi iz}} - \frac{a_-}{1-e^{2\pi iz}} + \hat{M}(z)\right)\Bigg|_{z=\frac{1}{2}-i\infty}^{z=\frac{1}{2}+i\infty} \end{aligned}$$

In particular the index of the operator $\overset{+}{\Theta} A_0(\omega,D_{n-1})$ *is given by the formula*

$$(11.21)\qquad \begin{aligned} \operatorname{ind}\,\Theta^+ A_0 = {} & \frac{1}{2\pi}\,\Delta\arg A_0(\omega,\xi_{n-1})\Big|_{\xi_{n-1}=+\infty}^{\xi_{n-1}=-\infty} + \\ & + \frac{1}{2\pi}\,\Delta\arg\det\,(ta_+ + (1-t)a_-)\Big|_{t=0}^{t=1}, \quad 0 \leq t \leq 1 . \end{aligned}$$

12. The factorization of matrices and the construction of regularizer.

In two previous sections we have constructed such regularizers of the operators $\overset{+}{\Theta} A_0$ that (11.2) holds where T_1, T_2 are compact operators. Now we shall construct more precise left and right regularizers of $\overset{+}{\Theta} A_0$ such that

(12.1) $R_1 \overset{+}{\Theta} A_0 = I + K_1$,

(12.2) $\overset{+}{\Theta} A_0 R_2 = I + K_2$,

where K_1, K_2 are operators of finite rank that is the images of $K_i (i=1,2)$ are finite dimensional. In the scalar case (see sections 5 and 9) the solution of pseudodifferential equation on the half-line or the solution of the corresponding Riemann-Hilbert problem (2.7) was based on the factorization of an elliptic symbol. A factorization is also possible for elliptic matrices for a fixed ξ'', but in general the factors in the factorization are discontinuous functions of ξ''.

At first we consider subclass of elliptic matrices for which the factors in the factorization depend smoothly on the parameters.

Theorem 12.1: (see [5],[6]) Let $A_0(\omega, \xi_{n-1})$ be the same matrix as in the sections 10 and 11. Let the condition (10.7) be fulfilled. Assume that for all $\omega \in S^{n-3}$ the operator $\overset{+}{\Theta} A_0(\omega, D_{n-1})$ has a unique inverse in $\overset{+}{H}_0(\mathbb{R}^1)$. Then the matrix $A_0(\omega, \xi_{n-1})$ has the following factorization

(12.3) $A_0(\omega, \xi_{n-1}) = A_-(\omega, \xi_{n-1}) A_+(\omega, \xi_{n-1})$,

where

(12.4) $A_+^{-1}(\omega, \xi_{n-1}) = b_+^{-1}(\xi_{n-1}) + c_+(\omega, \xi_{n-1})$,

(12.5) $A_-(\omega, \xi_{n-1}) = a_+ b_-^{-1}(\xi_{n-1}) + c_-(\omega, \xi_{n-1})$,

$b_\pm(\xi_{n-1})$ are the matrices in (10.8), and $c_+(\omega, \xi_{n-1} + i\tau)$ is analytic in $\xi_{n-1} + i\tau$ for $\tau > 0$, C^∞ with respect to $(\omega, \xi_{n-1} + i\tau)$ for $\tau \geq 0$ and satisfies the estimates

(12.6) $$|D_\omega^p D_{\zeta_{n-1}}^k c_+(\omega, \xi_{n-1} + i\tau)| \leq \frac{C_{pk}}{(1+|\xi_{n-1}|+|\tau|)^{\frac{1}{2}+\delta_0+k}},$$

$$\tau \geq 0, \zeta_{n-1} = \xi_{n-1} + i\tau, \; k \geq 0, \; |p| \geq 0, \; \delta_0 > 0.$$

Similarly the matrix $c_-(\omega,\xi_{n-1}+i\tau)$ is analytic in $\zeta_{n-1}=\xi_{n-1}+i\tau$ for $\tau<0$, of class C^∞ in $(\omega,\xi_{n-1}+i\tau)$ for $\tau\le 0$, and satisfies the estimates (12.6) for $\tau\le 0$. In addition

$$\text{(12.8)} \qquad \det A_+(\omega,\xi_{n-1}+i\tau)\neq 0 \quad \text{for} \quad \tau\ge 0,$$

$$\det A_-(\omega,\xi_{n-1}+i\tau)\neq 0 \quad \text{for} \quad \tau\le 0.$$

The conditions of the Theorem 12.1 are fulfilled for strongly elliptic matrix $A_0(\omega,\xi_{n-1})$, that is for such elliptic matrix $A_0(\omega,\xi_{n-1})$ that the matrix $A_R(\omega,\xi_{n-1}) = \frac{A_0+A_0^*}{2}$ is positive definite for all $\omega\in S^{n-3}$. For example, if the matrix $A_0(\omega,\xi_{n-1})$ is close to the unit matrix

$$|A_0(\omega,\xi_{n-1})-I|<\varepsilon \quad , \quad \varepsilon \text{ is small},$$

then it is a strongly elliptic matrix and so it has a factorization of the form (12.13).

Conversely if a matrix $A_0(\omega,\xi_{n-1})$ has a factorization of the form (12.3) then the equation $\Pi^+A_0(\omega,\xi_{n-1})\tilde{u}_+ = \tilde{g}_+$ is uniquely solvable in $\tilde{H}^+_{0,N}(\mathbb{R}^1)$ and its solution can be written in the following form:

$$\text{(12.10)} \qquad \tilde{u}_+ = A_+^{-1}\Pi^+A_-^{-1}\tilde{g}_+ \quad .$$

We now consider the general case of an elliptic matrix $A_0(\omega,\xi_{n-1})$ which satisfies the condition (10.7).

Theorem 12.2 (see [5], [6]):

There are an integer N and matrices $Q_N^\pm(\omega,\xi_{n-1})$, $P_{2N}(\omega,\xi_{n-1})$ such that

$$\text{(12.11)} \qquad A_0(\omega,\xi_{n-1}) = \Lambda_-^N(\xi_{n-1})\,\Lambda_+^N(\xi_{n-1})\,Q_N^-(\omega,\xi_{n-1})\,Q_N^+(\omega,\xi_{n-1})P_{2N}^{-1}(\omega,\xi_{n-1}),$$

where $Q_N(\omega,\xi_{n-1})$ have the same properties as the matrices $A_\pm(\omega,\xi_{n-1})$ in (12.4), (12.5) and $P_{2N}(\omega,\xi_{n-1})$ is polynomial matrix in ξ_{n-1} of degree $2N$, $P_{2N}(\omega,\xi_{n-1}) = \sum_{k=0}^{2N} p_k(\omega)\xi_{n-1}^k$,

$$\text{(12.12)} \qquad \det P_{2N}(\omega,\xi_{n-1})\neq 0, \quad p_{2N}(\omega) = I.$$

We shall use the Theorem 12.2 for constructing the left regularizer of the operator $\Pi^+ A_0(\omega,\xi_{n-1})$. Let $\tilde{R}_1$ denote the operator defined by

$$(12.13)\quad \tilde{R}_1\tilde{g}_+ = \Pi^+ \Lambda_+^{-N}\Lambda_-^{-N} P_{2N}(Q_N^+)^{-1}\Pi^+(Q_N^-)^{-1}\tilde{g}_+ \quad .$$

If we apply $\tilde{R}_1$ to the left in the equation $\Pi^+ A_0\tilde{u}_+ = \tilde{g}_+$ we obtain (see [5],[6])

$$(12.14)\quad \tilde{u}_+ = \sum_{k=1}^{2N} C_k^+(\omega,\xi_{n-1})\, w_k + \tilde{R}_1\tilde{g}_+ \ ,$$

where $C_k^+(\omega,\xi_{n-1})$ are some symbols and

$$(12.15)\quad w_k = \frac{1}{2\pi}\int_{-\infty}^{\infty} \xi_{n-1}^{k-1} P_{2N}^{-1}\tilde{u}_+ d\xi_{n-1},\ 1 \le k \le 2N \ .$$

We note that the operator $\sum_{k=1}^{2N} C_k^+(\omega,\xi_{n-1}) \frac{1}{2\pi}\int_{-\infty}^{\infty} \xi_{n-1}^{k-1} P_{2N}^{-1}\tilde{u}_+ \, d\xi_{n-1}$

is an operator of finite rank.

The right regularizer $\tilde{R}_2$ such that (12.2) holds can be constructed in a similar way by using the following factorization of the matrix $A_0(\omega,\xi_{n-1})$:

$$A_0(\omega,\xi_{n-1}) = \Lambda_-^N\Lambda_+^N(P_{2N}^{(1)})^{-1} Q_{N1}^- Q_{N1}^+$$

where $Q_{N1}^{\pm}$, $P_{2N}^{(1)}$ have the same properties as $Q_N^{\pm}, P_{2N}$ in (12.11).

13. General boundary problem for system of pseudodifferential equation on a half-line.

Let $A(\omega,\xi_{n-1})$ be the same matrix as in (10.3) and let the condition (10.12) be fulfilled. Then the operator $p'_+A(\omega,D_{n-1})$ is a Fredholm operator from $H^+_{t,N}(\mathbb{R}^1)$ into $H_{t-\alpha,N}(\mathbb{R}^1_+)$ for every $\omega \in S^{n-3}$. We denote the dimension of the kernel of the operator $p'_+A(\omega,D_{n-1})$ by $n_+(\omega)$ and the dimension of its cokernel by $n_-(\omega)$. By analogy with the scalar case, to make the boundary value problem on $\mathbb{R}^1_+$ correctly posed we add to the equation (10.3) at least max $n_-(\omega)$ potentials so as to eliminate the cokernel of the operator $p_+'A(\omega,D_{n-1})$ and we impose at least $\max_\omega n_+(\omega)$ boundary conditions at $x_{n-1} = 0$ so as to eliminate the kernel of $p_+'A(\omega,D_{n-1})$. Let m_+ and m_- be any integers such that $m_+ \geq \max n_+(\omega)$, $m_- \geq \max_\omega n_-(\omega)$ and $m_+-m_-=n_+(\omega)-n_-(\omega) = m$, where m is the index of $p_+'A(\omega,D_{n-1})$. Let $B_j(\xi')$, $C_k(\xi')$, $1 \leq j \leq m_+$, $1 \leq k \leq m_-$, be homogeneous p-dimensional vectors, C^∞ for $\xi' \neq 0$. Suppose that

(13.1) $\quad \mathrm{ord}_{\xi'}\, B_j(\xi') = \beta_j$, $\mathrm{ord}_{\xi'} C_k(\xi') = \gamma_k$, $\mathrm{Re}\,\beta_j < t - \frac{1}{2}$, $\mathrm{Re}\,\gamma_k < \alpha - t - \frac{1}{2}$

Let $E_{jk}(\xi'')$ be homogeneous functions of order $\beta_{jk} = \beta_j + \gamma_k - \alpha + 1$, C^∞ for $\xi'' \neq 0$.

Consider in $\mathbb{R}^1_+$ the following problem

$$(13.2) \quad p_+' A(\omega,D_{n-1})v_+(x_{n-1}) + \sum_{k=1}^{m_-} \rho_k\, p_+'\, C_k(\omega,D_{n-1})\, \delta(x_{n-1}) = g(x_{n-1}),$$

$$(13.3) \quad p''B_j(\omega,D_{n-1})v_+(x_{n-1}) + \sum_{k=1}^{m_-} E_{jk}(\omega)\, \rho_k = h_j,\ 1 \leq j \leq m_+,$$

where ρ_k, h_j are complex numbers, $v_+(x_{n-1}) \in H^+_{t,N}(\mathbb{R}^1_+)$, $g \in H_{t-\alpha,N}(\mathbb{R}^1)$.
We suppose that the following condition holds:

(13.4) <u>The system of equation</u> (13.2), (13.3) <u>is uniquely solvable for any $\omega \in S^{n-2}$</u>.

If the condition (13.4) is fulfilled then the methods of the section 12 gives possibility to construct the left and the right inverse of the operator (13.2) (13.3).

We shall consider the construction of the left inverse. The construction of the right inverse is similar. Let R be the operator similar to (12.13):

$$(13.5) \quad Rg = \Lambda_+^{-t-2N}\, P_{2N}(\omega,D_{n-1})(Q_N^+)^{-1}\theta^+ (Q_N^-)^{-1} \Lambda_-^{t-\alpha} (D_{n-1})\, lg$$

We note that the kernel of the operator R is zero. Indeed, let $Rg = 0$. Then $\theta^+(Q_N^-)^{-1} \Lambda_-^{t-\alpha} lg = 0$. Therefore $(Q_N^-)^{-1} \Lambda_-^{t-\alpha} lg = g_-$, where $g_- \in H_\delta^-(\mathbb{R}^1)$, $|\delta| < \frac{1}{2}$, and $lg = Q_N^- \Lambda_-^{-(t-\alpha)} g_-$ has the support in $\mathbb{R}^1_-$.

Thus $g = p_+' l g = 0$.

By applying $\Lambda_+^t R$ to the equation (13.2) we obtain (see [5], [6]):

$$(13.6)\quad \Lambda_-^N \Lambda_+^{t-N} v_+ = \sum_{k=1}^{2N} w_k C_{k1}^+(\omega, D_{n-1})\delta(x_{n-1}) - \sum_{k=1}^{m_-} \rho_k C_{k2}^+(\omega, D_{n-1})\delta(x_{n-1}) + $$

$$+ \Lambda_+^t Rg ,$$

where

$$(13.7)\quad w_k = p'' D_{n-1}^{k-1} P_{2N}^{-1} \Lambda_+^t v_+, \ 1 \le k \le 2N ,$$

and C_{ki}^+ are some symbols.

The system (13.6) is equivalent to the system (13.2) since the kernel of R is zero.

The equation $\Lambda_-^N \Lambda_+^{t-N} v_+ = \varphi_+$ has a solution $v_+ \in H_t^+(\mathbb{R}^1)$ if and only if $p'' \Lambda_-^{-k} \varphi_+ = 0$ for $1 \le k \le N$ and then $v_+ = \Lambda_+^{-t} \Theta^+ \Lambda_-^{-N} \varphi_+$. Therefore the equation (13.6) is equivalent to the equation

$$(13.8)\quad v_+ = \sum_{k=1}^{2N} w_k \Lambda_+^{-t} \Theta^+ \Lambda_-^{-N} \Lambda_+^N C_{k1}^+ \delta - \sum_{k=1}^{m_-} \rho_k \Lambda_+^{-t} \Theta^+ \Lambda_-^{-N} \Lambda_+^N C_{k2}^+ \delta +$$

$$+ \Lambda_+^{-t} \Theta^+ \Lambda_-^{-N} \Lambda_+^{N+t} Rg$$

with the following additional conditions

$$(13.9)\quad p'' \Lambda_-^{-k} \Big(\sum_{k=1}^{2N} w_k C_{k1}^+ \delta - \sum_{k=1}^{m_-} \rho_k C_{k2}^+ \delta + \Lambda_+^t Rg \Big) = 0, \ 1 \le k \le N .$$

Now we substitute (13.8) in the boundary conditions (13.3):

We obtain

$$(13.10)\quad \sum_{k=1}^{2N} w_k p'' B_j \Lambda_+^{-t} \Theta^+ \Lambda_-^{-N} \Lambda_+^N C_{k1}^+ \delta - \sum_{k=1}^{m_-} \rho_k (p'' B_j \Lambda_+^{-t} \Theta^+ \Lambda_-^{-N} \Lambda_+^N C_{k2}^+ \delta - E_{jk}(\omega)) =$$

$$= h_j - p'' B_j \Lambda_+^{-t} \Theta^+ \Lambda_-^{-N} \Lambda_+^{N+t} Rg , \ 1 \le j \le m_- .$$

Therefore the solution of the problem (13.2), (13.3) is reduced in an equivalent way to the solution of the system (13.8), (13.9), (13.10), where w_k, $1 \leq k \leq 2N$, have the form (13.7). In order to solve this system we apply to (13.8) the operator $p'' D_{n-1}^{j-1} P_{2N}^{-1} \Lambda_+^t$. Then we obtain

$$(13.11)\quad w_j = \sum_{k=1}^{2N} w_k \, p'' \, D_{n-1}^{j-1} P_{2N}^{-1} \, \Theta^+ \Lambda_-^{-N} \Lambda_+^N \, C_{k1}^+ \delta - \sum_{k=1}^{m_-} \rho_k p'' \, D_{n-1}^{j-1} \cdot$$

$$P_{2N}^{-1} \, \Theta^+ \Lambda_-^{-N} \cdot \Lambda_+^N \, C_{k2}^+ \, \delta + p'' D_{n-1}^{j-1} P_{2N}^{-1} \Theta^+ \Lambda_-^{-N} \Lambda_+^{N+t} Rg, \quad 1 \leq j \leq 2N .$$

Thus the solution of the problem (13.2). (13.3) is equivalent to the solution of the system (13.8), (13.9), (13.10), (13.11) ; where w_k, $1 \leq k \leq 2N$, are given by (13.7). Moreover the solution of the problem (13.2), (13.3) is reduced to the solution of the algebraic system (13.9), (13.10), (13.11), which is a system of $3N + m_+$ equations with $2N + m_-$ unknowns w_k, $1 \leq k \leq 2N$, ρ_j, $1 \leq j \leq m_-$. Suppose that the condition (13.4) is fulfilled. Then the system (13.9), (13.10), (13.11) has a unique solution. Indeed let $w_k^{(0)}$, $1 \leq k \leq 2N$, $\rho_j^{(0)}$, $1 \leq j \leq m_-$ be a nontrivial solution of the homogeneous system (13.9), (13.10), (13.11). Then $v^{(0)}(x_{n-1}), \rho_j^{(0)}, 1 \leq j \leq m_-$, where $v^{(0)}$ is given by (13.8) with $g = 0$, is a nontrivial solution of the homogeneous problem (13.2), (13.3) and this is a contradiction with the condition (13.4). We note that if not all $w_k^{(0)}$, $1 \leq k \leq 2N$, are zero then $v^{(0)}(x_{n-1})$ is nonzero as it follows from the homogeneous system (13.11), which is a consequence of the system (13.8).

Therefore the matrix $M(\omega)$ of the system (13.9), (13.10), (13.11) has a maximal rank for all $\omega \in S^{n-3}$. Thus there exists a matrix $L(\omega)$ which is the left inverse to $M(\omega)$. Applying $L(\omega)$ to the system (13.9), (13.10), (13.11), we find w_k, $1 \leq k \leq 2N$, ρ_k, $1 \leq k \leq m_-$. When w_k, ρ_k are known then the formula (13.8) gives the expression for $v_+(x_{n-1})$.

Thus we shall construct the left inverse to the operator defined by the equations (13.2), (13.3).

We note that the system (13.8) is a system of integral equations with a degenerated kernel and we have repeated above the usual procedure for the solution of such integral equations. In conclusion we note that the method of this section can be used for the construction of the parametrix for the general boundary problems for the system of elliptic pseudodifferential equations (see [5], section 6).

14. Necessary and sufficient conditions for the existence of the uniquely solvable boundary problem.

In this section we shall find out when the condition (13.4) is fulfilled. We shall formulate the following two simple lemmas:

Lemma 14.1: Let $A(\omega)$ be a continuous family of the Fredholm operators, $\omega \in S^{n-3}$, which map the Banach space B_1 on the Banach space B_2. Then there exists vectors $g_k \in B_2$, $1 \leq k \leq m_-$ such that the operator $\mathcal{A}(\omega)(u, c_1,\dots,c_{m_-}) = A(\omega)u + \sum_{k=1}^{m_-} c_k g_k$, which maps $B_1 \times \mathbb{C}^{m_-}$ on B_2, has no cokernel for all $\omega \in S^{n-3}$.

Lemma 14.2: Let $\mathcal{A}(\omega)$ be a continuous family of the Fredholm operator which have no cokernel for all $\omega \in S^{n-3}$. Then for each $\omega_0 \in S^{n-3}$ the kernel of $\mathcal{A}(\omega)$ has a continuous basis in $U(\omega_0) \subset S^{n-3}$, where $U(\omega_0)$ is some neighbourhood of ω_0.

It follows from the Lemma 14.1 that there exist symbols $C_k(\xi'',\xi_{n-1})$, $1 \leq k \leq m_-$, such that the system

$$(14.1)\quad p_+' A(\omega,D_{n-1})v_+(x_{n-1}) + \sum_{k=1}^{m_-} \rho_k p_+' C_k(\omega,D_{n-1})\,\delta(x_{n-1}) = g(x_{n-1})$$

has a solution $v_+ \in H^+_{t,N}$, $\rho_1,\dots,\rho_{m_-}$ for every $g \in H_{t-\alpha,N}(\mathbb{R}^1)$.

Now we shall find out when there exist symbols $B_j(\xi')$, $E_{jk}(\xi'')$, such that the problem (13.12), (13.13) is uniquely solvable.

Lemma 14.3: Suppose that the system (14.1) has a solution $(v_+(x_{n-1}),\rho_1,\dots\rho_{m_-})$

<u>for every</u> $\omega \in S^{n-3}$ and $g(x_{n-1}) \in H_{t-\alpha}(\mathbb{R}^1_+)$. <u>There exist boundary conditions</u>

$$(14.2)\quad p''B_j(\omega, D_{n-1})\, v_+(x_{n-1}) + \sum_{k=1}^{m_-} E_{jk}(\omega)\rho_k = h_j,\ 1 \le j \le m_+ ,$$

<u>such that the condition</u> (13.4) <u>is fulfilled if and only if there exist a continuous basis on</u> S^{n-3} <u>of the solutions of the homogeneous system (14.1).</u> It follows from the Lemma 14.2 that the kernel of the operator defined by (14.1) is a vector bundle. It follows from Lemma 14.3 that this bundle must be trivial bundle for the existence of (14.2) such that the condition (13.4) is fulfilled.

In order to formulate the conditions of the existence of a uniquely solvable problem (14.1), (14.2) in more invariant terms we need the definition of the index of a family of Fredholm operators. Let $\mathrm{Vect}(S^{n-3})$ be the set of all classes of isomorphic vector bundles on S^{n-3}. We denote by $\underline{m}$ the trivial bundle. Let $E-F$ be the formal difference of two vector bundles. Two differences $E-F$ and $E'-F'$ are called equivalent if there exists a vector bundle Q such that

$E \oplus F' \oplus Q \approx E' \oplus F \oplus Q$ where $\oplus$ the direct sum of the vector bundles and $\approx$ means the isomorphism of the vector bundles. The set of the classes of the equivalent formal differences will be denoted by $\mathcal{K}(S^{n-3})$. The set $\mathcal{K}(S^{n-3})$ is an abelian group if we define $(E-F) + (E'-F') = E \oplus E' - F \oplus F'$ and $-(E-F) = F - E$.

Each vector bundle E defines an element of $\mathcal{K}(S^{n-3})$ which we shall denote by $[E]$. It is obvious that an arbitrary element of $\mathcal{K}(S^{n-3})$ can be written as $[E] - [F]$, where E and F are vector bundles. Let $A(\omega)$ be a continuous family of Fredholm operators and let $\mathcal{A}(\omega)(u, c_1, \ldots, c_{m_-}) = A(\omega)\, u + \sum_{k=1}^{m_-} c_k g_k$ be the operator which has no cokernel for each $\omega \in S^{n-3}$ (see Lemma 14.1). Then the kernel of $\mathcal{A}(\omega)$ is a vector bundle (see Lemma 14.2).

The element of $\mathcal{K}(S^{n-3})$ equal to $[\ker \mathcal{A}(\omega)] - [\underline{m}_-]$ is called the index of the family $A(\omega)$ and it will be denoted by $\kappa(A(\omega))$.

It can be shown that $\kappa(A(\omega))$ does not depend on the choice of $g_1,\dots,g_{m_-}$ such that the operators $\mathcal{A}(\omega)$ maps $B_1 \times \mathbb{C}^{m_-}$ on B_2.

Theorem 14.1: For m_+ and m_- sufficiently large ($m_+ - m_- = m$ is equal to the index of $p'_+A(\omega, D_{n-1})$) there exist $B_j(\xi')$, $C_k(\xi')$, $E_{jk}(\xi'')$, $1 \leq k \leq m_+$, $1 \leq j \leq m_-$, such that the condition (13.4) is fulfilled if and only if the index of the family $p'_+A(\omega, D_{n-1})$ is equal to $[\underline{m}_+] - [\underline{m}_-] = [\underline{m}]$.

The next theorem which is based on the results of Atiyah and Bott find out when $\kappa(p'_+A(\omega, D_{n-1})) = [\underline{m}]$. Let I_N be the unit matrix in $\mathbb{C}^N$.

Theorem 14.2: In order to $\kappa(pA(\omega, D_{n-1})) = [\underline{m}]$ it is necessary and sufficient that for N sufficiently large the matrix $\left\| \begin{matrix} A_0(\xi'', \xi_{n-1}) & 0 \\ 0 & I_N \end{matrix} \right\|$ is homotopic to the matrix

$$\left\| \begin{matrix} \left(\dfrac{\xi_{n-1} + i|\xi''|}{\xi_{n-1} - i|\xi''|} \right)^m I_1 & 0 \\ 0 & I_{N+p-1} \end{matrix} \right\|$$

in the class of elliptic matrix satisfying the condition (10.7).

15. Generalized mixed problem in the half-space.

Consider in $\mathbb{R}^n_+$ an elliptic system of differential equations:

$$(15.5) \quad \sum_{j=1}^{p} \hat{L}_{ij}(D) u_j(x) = f_i(x), \; 1 \leq i \leq p, \; \operatorname{ord}_\xi L_{ij} = s_i + t_j ,$$

with mixed boundary conditions

$$(15.2) \quad p'_+\Big(\sum_{j=1}^{p} \hat{B}_{1j}(D)\, lu_j + \sum_{j=1}^{m_-} \hat{G}_{1ij}(D')(v_j(x'') \times \delta(x_{n-1}))\Big) = g_{1i}(x') ,$$

$$(15.3) \quad p'_-\Big(\sum_{j=1}^{p} \hat{B}_{2j}(D)\, lu_j + \sum_{j=1}^{m_-} \hat{G}_{2ij}(D')(v_j(x'') \times \delta(x_{n-1}))\Big) = g_{2i}(x'),$$

$1 \leq i \leq m$,

$$(15.4)\quad p'' \sum_{j=1}^{p} B^{-}_{ij}(D)\, lu_j = h_i(x''),\ 1 \le i \le m_+ ,$$

where the symbols have the following orders

$$\operatorname{ord}_\xi B_{kij}(\xi) = m_{ki} + t_j,\quad \operatorname{ord}_{\xi'} G_{kij}(\xi') = m_{ki} + \gamma_j ,$$

$2m = \sum_{i=1}^{p} s_i + \sum_{j=1}^{p} t_j$ is equal to the order of $\det\|L_{ij}(\xi)\|_{i,j=1}^{p}$,
$B^{-}_{ij}(\xi)$ are analytic with respect to ξ_n for $\operatorname{Im}\xi_n < 0$, and

$$s + N > \max_{i,k} \left(m_{ki} + \frac{1}{2}\right),\quad \operatorname{Re}\ \beta_i < s - 1,\quad \operatorname{Re}\ \gamma_j < - s + \frac{1}{2},$$

s is non integer if $s \le \max_{i,k} \left(m_{ki} + \frac{1}{2}\right)$.

We suppose that

$$f \in H_{s-s_i,N}(\mathbb{R}^n_+),\ g_{1i} \in H'_{s-m_{1i}-\frac{1}{2},N}(\mathbb{R}^{n-1}_+),\ g_{2i} \in H'_{s-m_{2i}-\frac{1}{2},N}(\mathbb{R}^{n-1}_-),$$

$$h_i(x'') \in H''_{s-\operatorname{Re}\beta_i-1}(\mathbb{R}^{n-2})$$

and we choose $u \in H_{s+t_j,N}(\mathbb{R}^n_+)$, $v_j \in H''_{s+\operatorname{Re}\gamma_j - \frac{1}{2}}(\mathbb{R}^{n-2})$. Let $\omega \in S^{n-3}$ be fixed and consider the following family of mixed problem in $\mathbb{R}^2_+$ connected with (15.1) - (15.4):

$$(15.5)\quad \sum_{j=1}^{p} L_{ij}(\omega, D_{n-1}, D_n)\, u_j(x_{n-1}, x_n) = f_i(x_{n-1}, x_n),\ 1 \le i \le p ,$$

$$(15.6)\quad p'_+\Big(\sum_{j=1}^{p} B_{1ij}(\omega, D_{n-1}, D_n)\, lu_j(x_{n-1}, x_n) + \sum_{j=1}^{m_-} v_j G_{1ij}(\omega, D_{n-1})\, \delta(x_{n-1})\Big) = g_{1i}(x_{n-1}),$$

$$(15.7)\quad p'_-\Big(\sum_{j=1}^{p} B_{2ij}(\omega, D_{n-1}, D_n)\, lu_j(x_{n-1}, x_n) + \sum_{j=1}^{m_-} v_j G_{2ij}(\omega, D_{n-1})\, \delta(x_{n-1})\Big) = g_{2i}(x_{n-1}),$$

$$1 \le i \le m ,$$

$$(15.8)\quad p'' \sum_{j=1}^{p} B^{-}_{ij}(\omega, D_{n-1}, D_n)\, lu_j(x_{n-1}, x_n) = h_i,\quad 1 \le i \le m_+,$$

where v_j, h_i are the complex numbers,

$u_j \in H_{s+t_j,N}(\mathbb{R}^2_+)$, $f_i \in H_{s-s_i,N}(\mathbb{R}^2_+)$, $g_{1i} \in H'_{s-m_{1i}-\frac{1}{2},N}(\mathbb{R}^1_+)$,

$g_{2i} \in H'_{s-m_{2i}-\frac{1}{2},N}(\mathbb{R}^1_-)$.

Suppose that the following condition is fulfilled:

(15.9) For given s and for arbitrary $\omega \in S^{n-3}$ the boundary problem (15.5) - (15.8) is uniquely solvable.

Then the following theorem holds:

Theorem 15.1: Let the condition (15.9) be fulfilled. Then the generalized mixed problem (15.1) - (15.4) is uniquely solvable.

To prove this theorem take the Fourier transform with respect to x'' and then we make the change of variables $x_{n-1} = \frac{y_{n-1}}{1+|\xi''|}$, $x_n = \frac{y_n}{1+|\xi''|}$. When we put $\xi_{n-1} = (1+|\xi''|)\eta_{n-1}$, $\xi_n = (1+|\xi''|)\eta_n$ we obtain that the operator corresponding to the problem (15.1) - (15.4) can be considered as a pseudodifferential operator in $\mathbb{R}^{n-2}$ with the operator valued symbol $\mathfrak{A}(\omega)$, which is the operator corresponding to the system (15.5) - (15.8) (see [6] for further details). Therefore the condition (15.9) gives that the symbol $\mathfrak{A}(\omega)$ is invertible and so that the problem (15.1)-(15.4) is uniquely solvable. Now we shall discuss the condition (15.9).

Suppose that the boundary operators $B_{1ij}(D)$ and $B_{2ij}(D)$ fulfil for each $\omega \in S^{n-3}$ the Shapiro-Lopatinskii condition. It follows from the Lemma 14.3 applied to the system of differential equations (15.5) that there exists a smooth basis $e_r(\xi',x_n) = (e_{1r}(\xi',x_n),\dots,e_{pr}(\xi',x_n))$, $(\xi' \neq 0)$ of the solutions of the system

$$(15.10)\qquad \sum_{j=1}^{p} L_{ij}(\xi',D_n)\,u_j(x_n) = 0\,,$$

which decrease for $x_n \to +\infty$.

Let $E_{kj}(\xi',\xi_n)$ be the Fourier transform of $e_{kj}(\xi',x_n)$, where we put $e_{kj} = 0$ for $x_n < 0$.

The general solution of the equation (15.5) has the following form:

$$(15.11)\quad u_i(x_{n-1},x_n) = p_+(u_i^{(0)} + \sum_{j=1}^{p} E_{ij}(\omega,D_{n-1},D_n)(w_k(x_{n-1})\times \delta(x_n))),$$

where $u_i^{(0)} = \sum_{j=1}^{p} L_{ij}^{(1)}(\omega,D_{n-1},D_n)\, lf_j,\ 1 \le i \le p,$

is a particular solution of (15.5), $\|L_{ij}^{(1)}(\xi)\|$ is the matrix inverse to $\|L_{ij}(\xi)\|$, $w_k(x_{n-1}) \in H'_{s-\frac{1}{2},N}(\mathbb{R}^1)$ are arbitrary functions.

If we substitute (15.11) in the boundary equations (15.6),(15.7),(15.8) we shall obtain after easy transformations similar to those in section 2 that the solution of the problem (15.5)- (15.8) is equivalent to the solution of some boundary problem (13.2), (13.3).

Therefore the condition (15.9) can be reformulated as a condition of the uniquely solvability of the problem (13.2),(13.3).

We note that if $u_i \in H_{s+t_i,N}(\mathbb{R}^n_+)$ and s is sufficiently large then the operator $p'_+A(\omega,D_{n-1})$ has no kernel, so that the mixed problem (15.1),(15.2), (15.3) is uniquely solvable without additional boundary conditions (15.4).

If s is negative and $|s|$ is sufficiently large then the operator $p'_+A(\omega,D_{n-1})$ has no cokernel. So that the mixed problem (15.1)-(15.4) is uniquely solvable without potentials $G_{kij}(D')(v_j \times \delta)$. It is obvious what is the formulation of the mixed problem for bounded domain G with a smooth boundary Γ, so we don't give it here.

In conclusion we shall mention two boundary problems the solution of which is similar to the solution of mixed problems, considered above.

The first is the transmission problem for two domains with smooth boundaries, which have a common part.

The second is the crack problem, where the crack is a smooth surface inside of domain. We suppose that the boundary of the crack is also smooth.

REFERENCES

1. Atiyah, M. K-theory. Harvard Univ. 1964.

2. Atiyah, M, Bott. R; The index problem for manifolds with boundary. Bombay Coll. Diff. Analysis Oxford Univ. Press. 1964. p. 175-186.

3. Boutet de Mouvell, Boundary problems for pseudodifferential operators, Acta Math. 126 N.1-2(1971), 11-51.

4. Eskin, G. The conjugacy problem for equations of principal type with two independent variables, Trans. Moscow Math Soc. 21(1970), 263-316.

5. ________, Boundary value problems and the parametrix for systems of elliptic pseudodifferential equations. Trans. Moscow Math. Soc. 28(1973), 74-115.

6. ________, Boundary problems for elliptic pseudodifferential operators, (in Russian), Nauka, Moscow, 1-231.

7. ________, Asymptotics of solutions of elliptic pseudo-differential operators with a small parameter, Soviet Math. Doklady 14(1973)N.4.

8. Dynin A. On the theory of pseudodifferential operators on a manifold with a boundary, Soviet Math. Dokl. 10(1969), 575-578.

9. Cordes H and Hermann E. Pseudo-differential operators on a half-line, J. Math. Mech. 18(1969), 893-908.

10. Čan Zui Ho and Eskin G. Boundary value problems for parabolic systems of pseudodifferential equations, Soviet Math. Dokl 12(1971), 739-743.

11. Fichera G. Sul problema della derivata obliqua e sul problema misto per l'equazone di Laplace, Boll. Un. Math Ital. 7(1952), 367-377.

12. Gohberg I, and Krein M. Systems of integral equations on a half-line with kernels depending on the difference of arguments, Amer. Math.Soc. Transl. (2)14(1960) 217-287.

13. Gohberg I. and Krupnik N. The algebra generated by the one-dimensional singular integral operators with piecewise continuous coefficients, Functional Anal. Appl. 4(1970), 193-201.

14. Gahov F. Boundary problems, Fizmatgiz, Moscow, 1958.

15. Hormander L. Linear Partial Differential Equations, Springer-Verlag, Berlin, 1963.

REFERENCES CONTD

16. Lienard A. Probleme plan de la derivée oblique dans la théorie du potentiel, Jour. Ecole Polit. 5-7(1938), 35-158 and 177-226.

17. Magenes E. and Stampacchia G. I problemi al contorno per le equazioni differenziali di tipo ellittico. Ann. Sc. Normale Sup. Pisa 7(1958), 247-357.

18. Miranda C. Su alcuni aspetti della teoria delle equazioni ellittiche, Bull. Soc. Math. France 86(1958), 331-354.

19. Nirenberg. L. Pseudodifferential operators, Proc. Symposia Pure Math. V. 16 Amer. Math. Soc. 1970, pp. 147-168.

20. Peetre. J. Mixed problems for higher order elliptic equations in two variables, I - Ann. Scnola Norm. Super Pisa, 15(1961), 337-353.

21. Shamir E. Mixed boundary value problems for elliptic equations in the plane, the L_p-theory, Ann Scuola Norm. Super 17(1963), 117-139.

23. ________, Elliptic systems of singular integral operators I - Trans Amer. Math. Soc. 127(1967), 107-124.

24. ________, Boundary value problems for elliptic convolution systems "pseudodifferential operators" CIME, Edizione Cremonese Roma, 1969.

24. ________, Regularity of mixed second order elliptic problems, Israel Journ. of Math. 6(1968), 150-168.

25. Stampacchia G. Problemi al contorno elliptici con dato discontinui, dotate de soluzione holderiane, Ann. Math . pure e appl.51(1960), 1-38.

26. Simonenko I, A new general method of investigating linear operator equations of singular integral equation type I,II, Izv. Akad, Nauk. SSSR ser. Mat. 29(1965), 567-586, 757-782(Russian)

27. Šubin M., Factorization of matrices depending on a parameter and elliptic equations in a half-space, Math. USSR Sb. 14(1971), 65-84.

28. Višik M. and Eskin G., Elliptic equations in convolution in a bounded domain and their applications, Russian Math. Surveys 22(1967), 303-332.

29. ________________, Normally solvable problems for elliptic systems of equations in convolution, Math. USSR Sb. 4(1967), 303-332.

30. ________________, The Sobolev-Slobodetsky spaces of variable order with weights norms and their applications to the mixed problem, Sibirski Math. Journ. v.9.N.5 (1968).

CENTRO INTERNAZIONALE MATEMATICO ESTIVO

(C.I.M.E.)

HYPOELLIPTICITE POUR DES OPERATEURS DIFFERENTIELS SUR DES GROUPES DE LIE NILPOTENTS

B. HELFFER

(Centre de Mathématiques de l'Ecole Polytechnique
91128 - Palaiseau - France)

Corso tenuto a Bressanone dal 16 al 24 giugno 1977

HYPOELLIPTICITE POUR DES OPERATEURS DIFFERENTIELS SUR DES GROUPES DE LIE NILPOTENTS.

par B. HELFFER

(Centre de Mathématiques de l'Ecole Polytechnique

91128 - Palaiseau - France)

On se propose de montrer dans cet exposé, comment les résultats de Boutet de Monvel - Grigis - Helffer [3] s'appliquent en particulier à l'étude de l'hypoellipticité pour des opérateurs différentiels invariants sur un groupe de Lie nilpotent de rang de nilpotence 2. Il s'agit essentiellement de réinterpréter les conditions nécessaires et suffisantes d'hypoellipticité données dans [3] en termes de représentations de groupe.

§ 0 Enoncé des résultats

On considère un groupe de Lie G de dimension n, simplement connexe, connexe, nilpotent, dont l'algèbre de Lie $\mathfrak{g}$ admet une décomposition de la forme :

$$\mathcal{G} = \mathcal{G}_1 \oplus \mathcal{G}_2$$

avec $$[\mathcal{G}_1, \mathcal{G}_1] \subset \mathcal{G}_2$$

$$[\mathcal{G}, \mathcal{G}_2] = 0$$

On suppose que $\mathcal{G}_2$ est de dimension ≥ 1 (le cas où $\mathcal{G}_2 = 0$ est évident)
On munit cette algèbre de Lie de la famille de dilatations suivante :

$$\delta_t(Y) = t^j Y \quad \text{pour } Y \text{ dans } \mathcal{G}^j \ (j = 1,2) \text{ et } t > 0.$$

On sait que l'application exponentielle, $\exp : \mathcal{G} \to G$, définit un difféomorphisme. On identifie $\mathcal{G}$ avec les champs de vecteurs réels invariants à gauche sur G en associant à X dans $\mathcal{G}$ le champ de vecteurs, encore noté X, défini par :

$$\forall \varphi \in C_o^\infty(G), \qquad (X\varphi)(x) = \frac{d}{dt} \varphi\Big(x.\exp(tX)\Big)/_{t=0}$$

Cette identification s'étend de manière unique en un isomorphisme entre l'algèbre enveloppante (complexifiée) $U(\mathcal{G})$ et l'algèbre de tous les opérateurs différentiels invariants à gauche sur G (à coefficients complexes).

Tout opérateur différentiel invariant à gauche sur G est de la forme :

$$(0.1) \qquad P = \sum_{|\alpha| \leq m} a_\alpha X_1^{\alpha_1} X_2^{\alpha_2} \ldots X_n^{\alpha_n}$$

avec $$a_\alpha \in \mathbb{C}, \qquad X_i \in \mathcal{G}$$

Si $(X_1,\ldots,X_n)$ forment une base ordonnée, le développement est unique.

Un opérateur différentiel invariant à gauche sur G est dit homogène d'ordre m si :

$$\delta_t(P) = t^m . P$$

Pour simplifier, dans cet exposé, nous ne donnerons un énoncé que dans le cas d'opérateurs différentiels homogènes. On a des théorèmes analogues dans le cas non-homogène (en remplaçant dans l'énoncé hypoelliptique par hypoelliptique avec perte de $m/2$ dérivées).

Dans la suite, on appellera groupe nilpotent de rang 2, un groupe ayant les propriétés ci-dessus.

On démontre le théorème suivant :

Théorème 0.1

Soit G un groupe nilpotent de rang 2, soit P un opérateur différentiel homogène de degré m sur G, alors P est hypoelliptique si, et seulement si, pour toute représentation π, unitaire, irréductible, non triviale, $\pi(P)$ est injectif dans $\mathcal{S}_\pi$ (où $\mathcal{S}_\pi$ désigne l'espace des vecteurs C^∞ de la représentation).

La formulation du théorème est due à Rockland [5], qui a démontré le théorème dans le cas du groupe de Heisenberg H_n. Dans [1], Beals montre que la condition du théorème est nécessaire pour tous les groupes nilpotents (munis d'une dilatation) et il démontre la condition suffisante pour les groupes $H_n \times \mathbb{R}^k$. Ultérieurement à notre travail, Beals donne dans [2] une nouvelle démonstration du théorème 0.1 en utilisant ses classes $S^{\Phi,\varphi}$.

§ 1. Rappels sur les résultats de [3]

Soit $P = p(x,D)$ un opérateur pseudo-différentiel sur $\mathbb{R}^n$ de degré m. Nous supposerons que P est régulier, i.e que son symbole total p admet un développement asymptotique.

(1.1) $$p(x,\xi) \sim \sum_{j=0}^{\infty} p_j(x,\xi)$$

où p_j est homogène de degré $m-j$ (j est un entier ou un demi-entier positif).

Soit Σ un cône lisse de codimension p dans $T^*\mathbb{R}^n \setminus 0$. Nous dirons que P est nul d'ordre k sur Σ et nous écrirons : $P \in \mathfrak{N}^{m,k}$ (ou $p(x,\xi) \in \mathfrak{N}^{m,k}$), si pour tout j inférieur ou égal à $k/2$, p_j s'annule à l'ordre $k-2j$.

Soit (x,ξ) un point de Σ, on définit, pour un élément P dans $\mathfrak{N}^{m,k}$,

(1.2) $$\sigma^k_{(x,\xi)}(P) = \sum_{|\alpha+\beta|+2j=k} \frac{1}{\alpha!\beta!} \left(\frac{\partial}{\partial x}\right)^{\alpha} \left(\frac{\partial}{\partial \xi}\right)^{\beta} p_j(x,\xi)\, y^{\alpha} D_y^{\beta}$$

$\sigma^k_{(x,\xi)}(P)$ est un élément de $A_n(\mathbb{C})$ (algèbre de Weyl des opérateurs différentiels à coefficients polynomiaux sur $\mathcal{S}(\mathbb{R}^n)$).

Pour un élément de $\mathfrak{N}^{m,o}$ (opérateurs pseudo-différentiels réguliers), $\sigma^o_{x,\xi}(P)$ est simplement le symbole principal de P au point (x,ξ).

On a les propriétés suivantes :

Pour P dans $\mathfrak{N}^{m,k}$ et Q dans $\mathfrak{N}^{m',k'}$

(1.3) $$\sigma^{k+k'}_{(x,\xi)}(P.Q) = \sigma^k_{(x,\xi)}(P) \,.\, \sigma^{k'}_{(x,\xi)}(Q)$$

(1.4) $$\sigma^{k+k'-2}_{(x,\xi)}\left([P,Q]\right) = \left[\sigma^k_{(x,\xi)}(P),\ \sigma^{k'}_{(x,\xi)}(Q)\right]$$

En particulier si $k = k' = 1$

(1.5) $$\sigma^o_{(x,\xi)}\left([P,Q]\right) = \frac{1}{i}\,\{p_o, q_o\}_{(x,\xi)}$$

Coordonnées adaptées

Au voisinage d'un point (x_o,ξ_o) de Σ, on introduit un système de coordonnées :

$$u = (u_1,..,u_p)$$
$$v = (v_1,..,v_{2n-p})$$

où les u_i (resp. v_i) sont C^∞, homogènes de degré 0 (resp. 1), non toutes nulles, de sorte que, au voisinage de (x_o,ξ_o), Σ est définie par le système d'équations : $u_i = 0 (i = 1,..,p)$. Soit U_j l'opérateur $u_j(x,D)$ (dans $\mathfrak{N}^{o,1}$), alors si P est dans $\mathfrak{N}^{m,k}$, on peut toujours écrire :

(1.6) $$P = \sum_{|\alpha|\leq k} A_\alpha \cdot U_1^{\alpha_1} \dots U_p^{\alpha_p}$$

où A_α est un opérateur pseudodifférentiel d'ordre $m-k/_2 + |\alpha|/2$

On a, en tout point (x,ξ) de Σ :

(1.7) $$\sigma^k_{(x,\xi)}(P) = \sum_{|\alpha|\leq k} \sigma^o_{x,\xi}(A_\alpha)\left(\sigma^1_{x,\xi}(U_1)\right)^{\alpha_1} \dots \left(\sigma^1_{x,\xi}(U_p)\right)^{\alpha_p}$$

Construction de parametrixes

On dira que P dans $\mathfrak{N}^{m,k}$ est transversalement elliptique de long de Σ, s'il est elliptique d'ordre m en dehors de Σ, et si tout point de Σ possède un voisinage conique Γ dans lequel on a : $\exists\, C > 0$, $\forall (x,\xi) \in \Gamma$, $|\xi| \geq 1$

(1.8) $$|p_o(x,\xi)| \geq C\, |\xi|^m \left|\sum_{i=1}^{p} u_i^2\right|^{k/2}$$

On a alors le théorème suivant (cf Th. 5.2.1 de [3])

Théorème 1.1

Soit P un opérateur pseudodifférentiel dans $\mathcal{N}^{m,k}$ transversalement elliptique le long de Σ, et soit (x_o,ξ_o) un point de Σ.

Les assertions suivantes sont équivalentes :

(i) Il existe un voisinage conique de (x_o,ξ_o) dans $T^*\mathbb{R}^n\backslash 0$, dans lequel P est microlocalement hypoelliptique avec perte de k/2 dérivées.

(ii) $\sigma^k_{(x_o,\xi_o)}(P)$ est injectif dans $\mathcal{S}'(\mathbb{R}^n)$.

Remarque 1.2 : Pour montrer l'hypoellipticité au voisinage de l'origine, il suffit de vérifier la condition aux points $(0,\xi)$ de Σ.

Etude de la condition (ii)

La clé de l'étude de la condition (ii) réside dans la forme particulière qu'a $\sigma^k_{(x,\xi)}(P)$ (cf. 1.7). $\sigma^k_{(x,\xi)}(P)$ ne dépend en fait que de p opérateurs différentiels du 1er ordre : $\sigma^1_{x,\xi}(U_i)$ $(i = 1, ..,p)$ que nous noterons dans la suite L^i_λ $(i = 1, ..,p)$ avec $\lambda = (x,\xi)$.

Avec ces notations, on écrit $\sigma^k_{(x,\xi)}(P)$ sous la forme :

$$(1.9) \qquad \sigma^k_{(x,\xi)}(P) = \mathcal{a}_{L,\lambda} \equiv \sum_{|\alpha|\leq k} a_\alpha(\lambda)\,(L^1_\lambda)^{\alpha_1} \ldots (L^p_\lambda)^{\alpha_p}$$

On considère $\mathcal{a}_{L,\lambda}$ comme un opérateur défini sur $\mathcal{S}'(\mathbb{R}^n)$.

λ est désormais <u>fixé</u>. On désigne par l^j_λ le symbole de L^j_λ $(j = 1,..p)$. La donnée des p l^j_λ $(j = 1,..,p)$ définit dans $\mathbb{R}^{2n^*}_{y,\eta}$ un p-plan que nous noterons $\mathcal{L}_\lambda$.

A ce plan p - plan $\mathcal{L}_\lambda$ est attaché un invariant symplectique qui est le rang de la 2 - forme symplectique sur $\mathbb{R}^{2n^*}$ restreint à $\mathcal{L}_\lambda$.

On pose :

(1.10) $$2\,q(\lambda) = \operatorname{rang}\left[\left\{l_\lambda^i, l_\lambda^j\right\}\right]$$

et on vérifie que :

(1.11) $$2\,q(\lambda) = \operatorname{rang}\left[\{u_i, u_j\}_{(\lambda)}\right]$$

$2\,q(\lambda)$ est donc le rang au point λ de la 2 - forme symplectique sur $T^*\mathbb{R}^n\backslash 0$ restreinte à Σ.

Par une transformation symplectique dans $\mathbb{R}^{2n}$, on peut trouver une forme canonique pour $\mathcal{L}_\lambda$ (en fonction de $2\,q(\lambda)$). Il existe un opé - rateur unitaire U_λ de $L^2(\mathbb{R}^n)$ (continu de $\mathcal{S}(\mathbb{R}^n)$ dans $\mathcal{S}(\mathbb{R}^n)$ et de $\mathcal{S}'(\mathbb{R}^n)$ dans $\mathcal{S}'(\mathbb{R}^n)$) associé à cette transformation symplectique tel que:

(1.12) $$U_\lambda^{-1}\, \mathcal{a}_{L,\lambda}\, U_\lambda = \mathcal{a}_\lambda (x,y,D_x)$$

On a pris ici comme nouvelles coordonnées sur $\mathbb{R}^n$; (x,y,z) avec

$x = x_1, \ldots, x_{q(\lambda)}$

$y = x_{q(\lambda)+1}, \ldots, x_{p-q(\lambda)}$

$z = x_{p-q(\lambda)}, \ldots, x_n$

On dira que $\mathcal{a}_\lambda(x,y,D_x)$ est une forme réduite de $\mathcal{a}_{L,\lambda}$. Dans sa forme réduite, on considèrera $\mathfrak{a}_\lambda(x,y,D_x)$ comme un opérateur sur $\mathcal{S}(\mathbb{R}^q)$ dépendant du paramètre y dans $\mathbb{R}^{p-2q(\lambda)}$.

Il résulta alors du théorème 3.1 de [3], la proposition suivante :

Proposition 1.3

Sous les hypothèses du Théorème 1.1, les deux conditions suivantes sont équivalentes :

(ii) $\sigma_\lambda^k(P)$ est injectif dans $\mathcal{S}'(\mathbb{R}^n)$

(iii) $\forall y \in \mathbb{R}^{p-2q(\lambda)}$, $\mathfrak{a}_\lambda(x,y,D_x)$ est injectif dans $\mathcal{S}(\mathbb{R}^{q(\lambda)})$.

En conclusion, nous venons de construire en tout point λ de Σ et pour tout y dans $\mathbb{R}^{p-2q(\lambda)}$ une application $\pi_{\lambda,y}$ de $\mathfrak{N}^{m,k}$ dans $A^{q(\lambda)}(\mathbb{C})$ définie par :

(1.13) $$\mathfrak{N}^{m,k} \ni P \to \pi_{\lambda,y}^k(P) = U_\lambda^{-1}\ \sigma_\lambda^k(P)\ U_\lambda = \mathfrak{a}_\lambda(x,y,D_x)$$

telle que :

si P est dans $\mathfrak{N}^{m,k}$ et Q est dans $\mathfrak{N}^{m',k'}$

(1.14) $$\pi_{\lambda,y}^{k+k'}(PQ) = \pi_{\lambda,y}^{k}(P).\,\pi_{\lambda,y}^{k'}(Q)$$

(1.15) $$\pi_{\lambda,y}^{k+k'-2}([P,Q]) = \left[\pi_{\lambda,y}^{k}(P),\ \pi_{\lambda,y}^{k'}(Q)\right]$$

On réécrit le théorème (1.1) sous la forme suivante :

Théorème 1.4

Soit P un opérateur pseudodifférentiel dans $\mathfrak{N}^{m,k}$, transversalement elliptique le long de Σ, et λ un point de Σ.

Les assertions suivantes sont équivalentes :

(i) Il existe un voisinage conique de λ dans $T^*\mathbb{R}^n\setminus 0$, dans lequel P est hypoelliptique avec perte de k/2 dérivées.

(ii) $\forall y \in \mathbb{R}^{p-2q(\lambda)}$, $\pi_{\lambda,y}(P)$ est injectif dans $\mathcal{S}(\mathbb{R}^{q(\lambda)})$.

§ 2. Application à l'étude de l'hypoellipticité pour des opérateurs différentiels invariants à gauche sur des groupes nilpotents de rang de nilpotence 2.

Soit $(X_1,.., X_p)$ une base de $\mathcal{G}_1$, $(Z_{p+1},.., Z_n)$ une base de $\mathcal{G}_2$ et soit P un opérateur différentiel invariant homogène d'ordre m.

$$(2.1) \qquad P = \sum_{|\alpha|+2|\beta|=m} a_{\alpha\beta}\, X_1^{\alpha_1}\dots X_p^{\alpha_p}\, Z_{p+1}^{\beta_{p+1}}\dots Z_n^{\beta_n}$$

Soit $u_j(x,\xi)$ le symbole principal de iX_j $(j=1,..,p)$ et soit Σ le cône lisse dans $T^*G\setminus 0$ défini par :

$u_1(x,\xi) = 0,.., u_p(x,\xi) = 0.$

Σ est un sous-fibré vectoriel de $T^*G\setminus 0$. On peut identifier G à $\mathbb{R}^n$ en prenant les coordonnées exponentielles; l'élément neutre du groupe est l'origine dans $\mathbb{R}^n$. On voit aisément que P défini par (2.1) est dans la classe $\mathfrak{N}^{m,m}(\mathbb{R}^n,\Sigma)$.

Remarquons enfin que $\Sigma \cap \{0\}\times\mathbb{R}^n$ s'identifie à $\mathcal{G}_1^\perp$ l'annulateur de $\mathcal{G}_1$ dans $\mathcal{G}^*$. On peut aussi l'identifier à $\mathcal{G}_2^*$.

Nous allons montrer que la condition du théorème (0.1) est suffisante, la nécessité résultant d'un théorème de Beals [1].

Ellipticité Transversale

soit $\xi \in \mathbb{R}^n \backslash 0$, on considère la représentation de $\mathcal{G}$ suivante :

$$\pi_\xi(X_j) = i\,\xi_j \qquad j = 1,..,p$$

$$\pi_\xi(Z_j) = 0 \qquad j = p+1,..,n$$

C'est une représentation scalaire de $\mathcal{G}$ qui correspond à la représentation $\tilde{\pi}_\xi$ de G suivante :

Si on choisit des coordonnées sur G (x,z) telles que :

$$G \in g = \exp\left(\sum_{j=1}^{p} x_j\, X_j + \sum_{j=p+1}^{n} z_j\, Z_j\right)$$

On a $\tilde{\pi}_\xi(g) = e^{i \sum_{j=1}^{p} x_j . \xi_j}$

$\tilde{\pi}_\xi$ est une représentation irréductible unitaire, non triviale si $\xi \neq 0$.

La condition d'injectivité de $\pi_\xi(P)$ s'écrit alors simplement :

(2.2) $$\sum_{|\alpha|=m} a_{\alpha,0}\, \xi^\alpha \neq 0 \qquad \text{pour } \xi \neq 0$$

C'est la condition d'ellipticité transversale pour P.

Si on identifie $\mathbb{R}^p$ et $\mathcal{G}_2^\perp$ l'annulateur de $\mathcal{G}_2$ dans $\mathcal{G}^*$, les représentations construites sont celles qui sont associées par la méthode des orbites (cf. [4] p 154) aux éléments de $\mathcal{G}_2^\perp$.

P est donc un élément de $\mathfrak{M}^{m,k}$ transversalement elliptique et nous allons appliquer le théorème (1.4).

Démonstration du théorème 0.1

On doit vérifier la condition du théorème (1.4) pour tout point λ

dans $\Sigma \cap \{0\} \times \mathbb{R}^n$ (autrement dit, λ dans $\mathcal{G}_1^{\perp}$). En effet l'hypoellipticité dans un voisinage de l'origine en résultera (Remarque 1.2), et, l'opérateur étant invariant à gauche, il sera hypoelliptique dans G.

On définit les représentations de $\mathcal{G}$ suivantes; pour λ dans $\mathcal{G}_1$ et y dans $\mathbb{R}^{p-2q(\lambda)}$; $\pi_{\lambda,y}$ est défini par :

$$X_i \to \pi^1_{\lambda,y}(X_i) \qquad i = 1,..,p$$

$$Z_i \to \pi^0_{\lambda,y}(Z_i) = \sigma(Z_i)(\lambda) \qquad i = p+1,..,n$$

On a simplement utilisé le fait que X_i appartient à $\mathfrak{N}^{1,1}$ (i=1,..,p) et que Z_i appartient à $\mathfrak{N}^{1,0}$ (i=p+1,..,n).

Le théorème (0.1) résulte de la vérification du point suivant :

$\pi_{\lambda,y}$ est associée à une représentation $\tilde{\pi}_{\lambda,y}$ irréductible, unitaire, non triviale de G.

On va construire explicitement $\tilde{\pi}_{\lambda,y}$.

Il résulte de l'étude faite au § 1 (et de [3]) que l'on peut trouver, λ étant fixé, une nouvelle base de $\mathcal{G}_1'$:

X_i $\left(i=1,..q(\lambda)\right)$, X_i' $\left(i=1,..,q(\lambda)\right)$, Y_i $\left(i=1,..,p-2q(\lambda)\right)$

telle que :

$$\pi_{\lambda,y}(X_j) = i\,x_j \qquad (j=1,..,q(\lambda)$$

$$\pi_{\lambda,y}(X_j') = \frac{\partial}{\partial x_j} \qquad (j=1,..,q(y)$$

$$\pi_{\lambda,y}(Y_j) = i\,y_j \qquad (j=1,..,p-2q(\lambda)$$

et

$$\pi_{\lambda,y}(Z_j) = \sigma(Z_j)(\lambda) \overset{\text{def}}{\equiv} i\,z_j(\lambda) \qquad (j=p+1,..,n-2p)$$

Soient $(\tilde{x},\tilde{x}',\tilde{y},\tilde{z})$ les coordonnées exponentielles sur G associées à la base choisie ci-dessus :

$$G \ni g = \exp\left(\sum_{j=1}^{q(\lambda)} \tilde{x}_j X_j + \sum_{j=1}^{q(\lambda)} \tilde{x}'_j X'_j + \sum_{j=1}^{p-2q(\lambda)} \tilde{y}_j . Y_j + \sum_{j=p+1}^{n} \tilde{z}_j . Z_j\right)$$

On pose pour v dans $L^2(\mathbb{R}^{q(\lambda)})$

$$\tilde{\pi}_{\lambda,y}(g).v(x) = e^{i\left[\sum_{j=1}^{p-2q(\lambda)} \tilde{y}_j . y_j + \sum_{j=1}^{q(\lambda)} \tilde{x}_j . x_j + \sum_{j=p+1}^{n} \tilde{z}_j . z_j(\lambda) + \sum_{j=1}^{q(\lambda)} \frac{\tilde{x}_j \tilde{x}'_j}{2}\right]} \times v(x + \tilde{x}')$$

On vérifie que c'est une représentation de G dans $L^2(\mathbb{R}^{q(\lambda)})$en utilisant la formule de Campbell-Hausdorff pour les groupes nilpotents de rang 2 : $e^X . e^Y = e^{X+Y+\frac{[X,Y]}{2}}$

Cette représentation est unitaire; elle donne bien sur $\mathfrak{G}$ la représentation $\pi_{\lambda,y}$; elle est non triviale car $\lambda \neq 0$. On montre facilement qu'elle est irréductible (cf. [4], p 72). Le théorème (0.1) est ainsi complètement démontré.

<u>Remarque 2.1</u> : <u>lien avec la méthode des orbites</u> (cf. [4])

On désigne par $\mathfrak{G}_1^0$ (resp. $\mathfrak{G}_1^1$) l'espace vectoriel engendré par les Y_j $(j=1,.., p-2q(\lambda))$ (resp. les X'_j $(j=1,.., q(\lambda))$). On pose

$\mathfrak{G}^0 = \mathfrak{G}_1^1 \oplus \mathfrak{G}_1^0 \oplus \mathfrak{G}_2$.

Soit $l^*_{\lambda,y}$ la forme linéaire sur $\mathfrak{G}$ associé à (λ,y) de la manière suivante :

$$1^{*}_{\lambda,y}\left(\sum_{j=1}^{q(\lambda)}\tilde{x}_jX_j+\sum_{j=1}^{q(\lambda)}x'_jX'_j+\sum_{j=1}^{p-2q(\lambda)}\tilde{y}_j.Y_j+\sum_{j=p+1}^{n}\tilde{z}_jZ_j\right)=$$

$$=\sum_{j=p+1}^{n}\tilde{z}_j.z_j(\lambda)+\sum_{j=1}^{p-2q(\lambda)}\tilde{y}_j.y_j$$

A l'aide du théorème de [4] p 154, on peut vérifier que, si G^o est le groupe associé à $\mathcal{G}_o$, la représentation $\tilde{\pi}_{\lambda,y}$ est équivalente à la représentation induite $T(\mathcal{G}_o, 1^{*}_{\lambda,y}) = \operatorname{ind}_{G_o \nearrow G} \chi_{(\lambda,y)}$

où $\chi_{\lambda,y}$ est la représentation $\chi_{\lambda,y}(\exp X) = \exp i\,(1^{*}_{y,\lambda}, X)$ pour X dans $\mathcal{G}_o$. On peut ainsi montrer qu'on a obtenu avec les représentations $\tilde{\pi}_{\xi}$ et $\tilde{\pi}_{\lambda,y}$ toutes les classes d'équivalence de représentations irréductibles, si l'on remarque que deux éléments 1^{*} et $1'^{*}$ qui coincident sur $\mathcal{G}_2$ sont sur la même orbite de la représentation coadjointe de G, si et seulement si, elles coincident sur $\mathcal{G}_1^o$.

BIBLIOGRAPHIE

[1] R. Beals - Séminaire Goulaouic - Schwartz (1977)

[2] R. Beals - Comptes rendus du colloque de St-Jean de Monts (Juin I977)

[3] L. Boutet de Monvel, A. Grigis, B. Helffer - Parametrixes d'opérateurs pseudodifférentiels à caractéristiques multiples Astérisque 43-35 (1976), p 93 - 121

[4] L. Pukanszky - Leçon sur les représentations des groupes Dunod (1967)

[5] C. Rockland - Hypoellipticity on the Heisenberg.Group. Representation Theoretic criteria, (preprint)

[6] L.P. Rothschild and E.M. Stein - Hypoelliptic differential operators and nilpotent groups. Acta Math. (1977).

CENTRO INTERNAZIONALE MATEMATICO ESTIVO

(C.I.M.E.)

LECTURES ON DEGENERATE ELLIPTIC PROBLEMS

J. J. KOHN

Princeton University - Princeton NJ,U.S.A.

Corso tenuto a Bressanone dal 16 al 24 giugno 1977

Lecture 1.

These lectures will be concerned with equations (and systems of equations) which are "close" to elliptic. By this we mean that they are limits of elliptic ones; such as, for example, the heat equation can be expressed as the following limit:

$$(1) \qquad -\frac{\partial^2}{\partial x^2} + \frac{\partial}{\partial t} = \lim_{\delta \to 0} \left(-\frac{\partial^2}{\partial x^2} - \delta \frac{\partial^2}{\partial t^2} + \frac{\partial}{\partial t}\right).$$

We will call this phenomenon degenerate ellipticity. The emphasis here will be on L_2-methods, we will be studying our equations by means of the following variational problem. Let Ω be a domain in $\mathbb{R}^n$ and let $\mathcal{D}$ be a set of m-tuples of functions on Ω. Let Q be a quadratic form on $\mathcal{D}$ given by:

$$(2) \qquad Q(u, v) = \sum_{i,j} \sum_{|\alpha|, |\beta| \le m} (a_{\alpha\beta ij} D^\alpha u^i, D^\beta v^j),$$

where

$$u = (u^1, \dots, u^m), \quad v = (v^1, \dots, v^m)$$

and $(\ ,\ )$ denotes the L_2-inner product on Ω. Given an m-tuple $f = (f', \dots, f^m)$ of functions on Ω, we associate with Q the following variational problem, to find $u \in \mathcal{D}$ such that we have

$$(3) \qquad Q(u, v) = (f, v) \text{ for all } v \in \mathcal{D}.$$

Let P be the map of $\mathcal{D}$ into m-tuples on Ω defined by:

(4) $$(Pu)_j = \sum_i \sum_{|\alpha|, |\beta| \leq m} D^\beta(a_{\alpha\beta ij} D^\alpha u^i), \quad j = 1, \ldots, m$$

then, if $\mathcal{D}$ contains all m-tuples with compact support we have

(5) $$Pu = f,$$

whenever u satisfies (3).

In this way, the above variational problem can be used to study the operator P on the space $\mathcal{D}$. In many applications it is also useful to set $Q(u, v) = (Lu, Lv)$ where L is an operator of order m; in this case $P = L'L$, where L' denotes the formal adjoint of L, and our variational problem can be used to study the operators L and L'.

The following is a standard result in Hilbert space theory.

6. Theorem. Suppose Q satisfies the conditions:

(7) $$Q(u, v) = \overline{Q(v, u)}$$

(8) $$|Q(u, v)|^2 \leq Q(u, u)Q(v, v)$$

There exists $c > 0$ such that:

(9) $$\|u\|^2 \leq CQ(u, u) \quad \text{for all } u \in \mathcal{D}.$$

Further, if $\mathcal{D}$ is dense in the space $(L_2(\Omega))^m$, then for each $f \in (L_2(\Omega))^m$ there exists a unique u in closure of $\mathcal{D}$ under the norm Q, such that

(10) $$Q(u, v) = (f, v) \quad \text{for all} \quad v \in \mathcal{D}.$$

Proof: Let $\tilde{\mathcal{D}}$ denote the closure of $\mathcal{D}$ under Q. Then the functional which sends $v \in \tilde{\mathcal{D}}$ into (f, v) is bounded in the Q-norm, since by (9):

$$|(f, v)| \leq \|f\| \, \|v\| \leq \sqrt{c} \, \|f\| \, \sqrt{Q(v, v)} \; .$$

Hence this functional has a unique representative $u \in \tilde{\mathcal{D}}$ which satisfies (10).

It is shown in [1] that the hypotheses (7) and (8) can be weakened considerably, thus giving this method mo re flexibility; here we will not enter into these matters since we want to concentrate on the main ideas.

Our attention in these lectures will be focussed on the smoothness of the solution u. For this purpose we will need a-priori estimates which are stronger then (9). To illustrate the method which will be developed we will give a proof of the hypoellipticity of the heat equation. This is probably the most difficult way of studying the heat equation but it has the advantage that it can be used to analyze much more general equations. Let L be the differential operator defined on function in $\mathbb{R}^2$ by:

(11) $$Lu = - u_{xx} + u_t$$

12. Theorem. If $u, f \in L_2(\Omega)$, $\Omega \subset \mathbb{R}^2$ and if u is a weak solution of the equation $Lu = f$ and if f is C^∞ or an open subset U of Ω then u is C^∞ on U.

Let $\zeta, \zeta' \in C_0^\infty(\Omega)$ such that $\zeta' = 1$ on a neighborhood of the support of ζ. Let $u \in C^\infty(\Omega)$ then from (11) we have

(12) $$(Lu, \zeta^2 u) = - (u_{xx}, \zeta^2 u) + (u_t, \zeta^2 u)$$

by integration by parts we obtain:

(13) $$(u_t, \zeta^2 u) = -2(u, \zeta\zeta_t u) - (u, \zeta^2 u_t)$$

hence

(14) $$(u_t, \zeta^2 u) = -(u, \zeta\zeta_t u)$$

and so

(15) $$|(u_t, \zeta^2 u)| \leq \text{const.}\ \|\zeta' u\|^2$$

from (12) we deduce

(16) $$\|\zeta u_x\|^2 \leq (\zeta Lu, \zeta u) + \text{const.}\ \|\zeta' u\|^2 .$$

Substituting $D_x^{k-1} u$ for u, with $k \geq 2$, and integrating by parts, we have:

(17) $$\|\zeta D_x^k u\|^2 \leq (\zeta D_x^{k-2} Lu, \zeta D_x^k u) + \text{const.}\ \|\zeta' D_x^{k-1} u\|^2 .$$

and using the Schwarz inequality we get:

(18) $$\|\zeta D_x^k u\|^2 \leq \text{const.}\ (\|\zeta D_x^{k-2} Lu\|^2 + \|\zeta' D_x^{k-1} u\|^2) .$$

Applying (18) with ζ replaced by ζ' and k by $k - 1$ to the last term of (18) we have:

(19) $$\|\zeta D_x^k u\|^2 \leq \text{const.}\ (\|\zeta D_x^{k-2} Lu\|^2 + \|\zeta' D_x^{k-3} Lu\|^2 + \|\zeta'' D_x^{k-2} u\|^2)$$

where $\zeta'' \in C_0^\infty(\Omega)$ and $\zeta'' = 1$ on a neighborhood of the support of ζ'. Thus, by induction and by choosing an increasing sequence of $\zeta_j \in C_0^\infty(\Omega)$ with $\zeta_1 = \zeta$ and $\zeta_k = \zeta'$ and $\zeta_j = 1$ on a neighborhood of the support of ζ_{j-1}

we obtain

(20) $$\|\zeta D_x^k u\|^2 \le \text{const.}\ (\sum_{j=0}^{k-2} \|\zeta' D_x^j Lu\|^2 + \|\zeta' u\|^2), \qquad k \ge 2$$

and the case $k = 1$ is estimated by (16). From (11) we obtain

(21) $$\zeta u_t = \zeta Lu + \zeta u_{xx}\ .$$

Applying D_x^{k-2} to this equation combining with (20) we deduce inductively that:

(22) $$\|\zeta D_x^{k-2} u_t\|^2 \le \text{const.}\ (\sum_{j=0}^{k-2} \|\zeta' D_x^j Lu\|^2 + \|\zeta' u\|^2)\ .$$

Finally, differentiating (21) above with respect to t and combining with the above, we obtain:

(23) $$\|\zeta D_x^k D_t^m u\|^2 \le \text{const.}\ (\sum_{\substack{i \le k-2 \\ j \le m-1}} \|\zeta' D_x^i D_t^j Lu\|^2 + \|\zeta' u\|^2)\ .$$

Up to now we have assumed that $u \in C^\infty$, and obtained the estimates (23) based on that assumption. Now assume that u is a weak solution of $Lu = f$ and that f is C^∞ in an open set U. We want to show that if $P \in U$ then u is C^∞ a neighborhood of P (and hence in all of U). So choose the function ζ, $\zeta' \in C_0^\infty$ so that $\zeta = 1$ in a neighborhood of P and $\zeta' = 1$ in a neighborhood of the support of ζ. Let $\varphi \in C_0^\infty(\mathbb{R}^n)$ with

(24) $$\int_{\mathbb{R}^n} \varphi(x) dx_1 \dots dx_n = 1$$

For $u \in L_2$ define $S_\delta u$ by:

(25)
$$S_\delta u(x) = \int_{\mathbb{R}^n} u(x+\delta y)\varphi(y)dy_1 \dots dy_n$$
$$= \delta^{-n} \int_{\mathbb{R}^n} u(y)\varphi(\frac{y-x}{\delta})dy_1 \dots dy_n \, .$$

Then, for $\delta \neq 0$, we have $S_\delta u \in C^\infty(\mathbb{R}^n)$. Furthermore we have

(26)
$$\zeta' D^\alpha L S_\delta u = \zeta' D^\alpha S_\delta f$$

and this converges to $\zeta' D^\alpha f$ for every α (since f is in C^∞ in a neighborhood of the support of ζ'. Hence, replacing u with $S_\delta u$ in (23) we conclude that for each α, $\|\zeta D^\alpha S_\delta u\|$ is bounded independently of δ. Therefore, $\zeta D^\alpha u \in L_2$ for all α and so u is in C^∞ on the interior of the set defined by $\zeta = 1$. Which concludes the proof of Theorem 12.

Now we will formulate a Dirichlet problem for the heat equation which will serve as a model for the more general degenerate elliptic case. Let Ω be a bounded region in $\mathbb{R}^2$ with a smooth boundary $b\Omega$, let r be a C^∞ function defined in a neighborhood of $b\Omega$ such that, $dr \neq 0$, $r > 0$ outside of $\overline{\Omega}$ and $r < 0$ in Ω. Then if $u \in C^\infty(\overline{\Omega})$ we have

(27)
$$(Lu, u) = \|u_x\|^2 + (u_t, u) + \int_{b\Omega} r_x u_x u \, ds$$

and since

$$(u_t, u) = -(u, u_t) + \int_{b\Omega} r_t u^2 \, ds$$

we have

(28)
$$(u_t, u) = \frac{1}{2} \int_{b\Omega} r_t u^2 \, ds$$

hence

(29) $$(Lu, u) = \|u_x\|^2 + \frac{1}{2}\int_{b\Omega} r_t u^2 \, ds + \int_{b\Omega} r_x u_x u \, ds .$$

Following Fichera (see [2]), we divide $b\Omega$ into three subsets Σ_1, Σ_2 and Σ_3. Σ_3 consists of those points where $r_x \neq 0$, Σ_2 is the subset of $b\Omega - \Sigma_3$ in which $r_t < 0$ and Σ_1 is the remainder (i.e. Σ_1 is given by $r_x = 0$ and $r_t \geq 0$). If we set $u = 0$ on the interior of $\Sigma_2 \cup \Sigma_3$ we have

(30) $$(Lu, u) \geq \|u_x\|^2$$

from this we can deduce that if $Lu = 0$ then $u = 0$ since u is constant on the line segments, $t = \text{const.}$ and dense set of these segments have end points in Σ_3. Thus, given $f \in C^\infty(\bar{\Omega})$ and v in C^∞ defined on the interior of $\Sigma_3 \cup \Sigma_2$ there exists at most one smooth function u on $\bar{\Omega}$ such that $Lu = f$ and $u = v$ on the interior of $\Sigma_3 \cup \Sigma_2$.

Suppose $Lu = 0$ and $u = v$ on $b\Omega$ and suppose $u \in C^\infty(\bar{\Omega})$. This implies that at certain points of $b\Omega$, v may have to satisfy compatibility conditions. To see this suppose that the origin is in $b\Omega$ and that r is of the form $r(x, t) = t - g(x)$ further let's suppose that

(31) $$g(x) = -Mx^2 + \text{higher order terms.}$$

The boundary near the origin is "flattened" by choosing the local coordinates

(32) $$\tau = t - g(x)$$
$$\xi = x .$$

Then

(33) $$Lu = -u_{\xi\xi} + 2g'u_{\xi\tau} - u_{\tau\tau}(g')^2 + (1+g'')u_\tau = 0\,,$$

here the prime denotes differentiation of $g(\xi)$ with respect to ξ. Since $u(\xi, 0) = v(\xi)$ and $g'(0) = 0$, we see that at the origin

(34) $$(1-2M)u_\tau(0) = v_{\xi\xi}(0)\,.$$

Thus we obtain as a necessary condition

(35) $$v_{\xi\xi}(0) = 0 \quad \text{in case} \quad M = \frac{1}{2}\,.$$

Now suppose $M \neq \frac{1}{2}$; differentiating (33) with respect to ξ we find, with the aid of (34), that at the origin

(36) $$v_{\xi\xi\xi} - (1-6M)u_{\xi\tau} - \frac{g'''v_{\xi\xi}}{1-2M} = 0\,.$$

Hence we obtain the compatibility condition: at the origin

(37) $$(1-2M)v_{\xi\xi\xi}(0) = g'''(0)v_{\xi\xi}(0) \quad \text{if} \quad M = \frac{1}{6}\,.$$

Next, assuming $M \neq \frac{1}{2}, \frac{1}{6}$, so that u_τ and $u_{\tau\xi}$ are known at the origin (from (34) and (36)), we differentiate (33), in turn, with respect to τ and twice with respect to ξ, and find that at the origin the following expressions are determined, in terms of derivatives of v:

(38) $$(2M-1)u_{\tau\tau} + u_{\tau\xi\xi}\,,$$
$$8M^2u_{\tau\tau} + (10M-1)u_{\tau\xi\xi}\,.$$

Thus again we shall obtain a compatibility condition on the derivatives of v at the origin in case the determinant of the coefficients vanishes, i. e.

$$12M^2 - 12M + 1 = 0 . \tag{39}$$

The general situation is given in the following theorem, which is proved in [1].

<u>Theorem</u> 36. Let v be the boundary values of a C^∞ solution of the heat equation in $t \geq -Mx^2 + \ldots$ (in a neighborhood of the origin). There is a sequence $M_1, M_2, \ldots$ of positive numbers such that, if M is equal to one of these values, then v and its derivatives satisfy necessary compatibility conditions at the origin. For any other value of M there are no compatibility conditions. The distinguished values M_i are obtained as follows: the numbers $c_i = \frac{1}{4M_i}$ are roots of the Lequerre polynomials $L_k^{(\alpha)}$ with $\alpha = \pm\frac{1}{2}$, k = 1, 2, ... These polynomials are defined by

$$L_k^{(\alpha)}(c) = \sum_{j=0}^{k} \binom{k+\alpha}{k-j} \frac{(-c)^j}{j!} . \tag{40}$$

Lecture 2.

The "physical" reason behind the compatibility conditions for the heat equation can be found in the uniqueness of solutions given by (26). Take the region Ω to be bounded by the a segment of the x-axis and the graph of the formation $t = g(x)$ where $g(x) < 0$ in the interior of the segment and 0 at the end points. Then the segment is in Σ_1, the priority $(x, g(x))$ with $g'(x) \neq 0$ are in Σ_3 an the rest of the boundary is in Σ_2.

Now translating let us place the origin at a point with $g(x)$ of the form (31). Then the value of u at the origin is completely determined by the values of u on $(x, g(x))$ with $x \leq 0$ and $g(x) \leq 0$ and also by the values of u or the points with $x \geq 0$, $g(x) \leq 0$. This is seen by looking at the Dirichlet problem in the regions $\{(x,t) | g(x) \leq t \leq 0, \ x \leq 0\}$ and $\{(x,t) | g(x) \leq t \leq 0, \ x \geq 0\}$. Thus, one should expect to have compatibility conditions.

We will now give a rough description of the results contained in [1]. Consider the equation:

(38) $$Lu = a^{ij} u_{ij} + b^i u_i + cu = f$$

with $\Sigma\, a^{ij} \xi_i \xi_j \geq 0$.

Here $u(x)$ is a real function defined in a compact domain $\Omega \subset \mathbb{R}^n$ (or on a manifold), with C^∞ boundary; $x = (x_1, \ldots, x_m)$ represent the coordinates, and we have used subscripts to denote differentiation; we have also used the summation convention. The coefficients are real and of class C^∞ in $\bar{\Omega}$. The first boundary value problem consists of prescribing the values of u or a certain portion of the boundary $b\Omega$. We wish to obtain unique solutions of the problem which are smooth up to and including the boundary. If the leading part is elliptic, we have the usual Dirichlet problem otherwise we proceed following the definitions of Fichera in [2]. Let $v = (v_1, \ldots, v_n)$ denote the exterior normal vector at point of $b\Omega$, we divide $b\Omega$ into three portions as follows:

Σ_3 = the set of non-characteristic boundary points, i.e., those where $a^{ij} v_i v_j > 0$,

Σ_2 = the set of characteristic boundary points for which $(b^i - a^{ij}_j)\nu_i > 0$,

$$\Sigma_1 = b\Omega - \Sigma_3 - \Sigma_2 .$$

The "Dirichlet problem" is that of finding a solution of (38) with prescribed values on $\Sigma_2 \cup \Sigma_3$. After subtraction of a function with the same values, we may suppose that the given boundary values on $\Sigma_2 \cup \Sigma_3$ are zero. We now improve the following conditions:

(a) $\Sigma_2 \cup \Sigma_3$ is closed.

(b) $-c$ is very large positive compared to the other coefficients and their derivatives of second order, and to the derivatives of order ≤ 3 of the r^{ij}, as well as to γ^{-1} defined below.

To formulate (c) we define certain invariants. Let x_0 be a point of Σ_2 and let r be a function which vanishes identically on $b\Omega$ with $dr \neq 0$ and $r < 0$ in Ω ; then, on Σ_2 the values of

$$\text{(39)} \qquad \beta = (b^i - a^{ij}_j) r_i, \qquad \alpha = (b^i - a^{ij}_j)(r_k r_m a^{km})_i$$

are independent of the particular coordinate system. By hypothesis, $\beta > 0$ on Σ_2 and, since the vector r_i is parallel to ν_i and $\nu_i\ \partial/\partial x_i (r_k r_m a^{km}) \leq 0$, it follows that $\alpha \leq 0$. Furthermore, α/β^2 is invariant under change of the function r or under multiplication of the operator L by a positive factor. The third condition is:

$$\text{(c)} \qquad \gamma = 1 + m \frac{\alpha}{\beta^2} > 0 \text{ on } \Sigma_2 .$$

To formulate the last condition set

$$\text{(40)} \qquad a = a^{km} r_k r_m ,$$

then a vanishes on Σ_2. Let $L_0 = a^{ij} \frac{\partial^2}{\partial x_i \partial x_j}$ then:

(d) For e constant ε_0 depending only on n we require

(41) $$\frac{\sqrt{L_0 a}}{\beta}(x_0) \leq \varepsilon_0 \frac{\gamma(x_0)}{2m} .$$

42. Theorem. If $f \in H_{2m}(\Omega)$ (i.e. all derivatives of f up to order 2m are square integrable) and if conditions (a), (b), (c) and (d) are satisfied with $\varepsilon_0 = (100n^2 3^n)^{-1}$ then the unique weak solution of (38) is in H_m.

The proof of this theorem is rather intricate and most of [3] is devoted to it. Here we will only make a few remarks about this theorem. First, the choice of ε_0 is much too small and u is much smoother then just in H_m; however, to describe the smoothness of u precisely we would have to discuss some rather complicated weight functions--which would take us too far afield. Second, (b) can very often be bypassed by changing the unknown function; for example the equation $u_{xx} - u_t = f$ is equivalent to the equation $v_{xx} - v_t - \lambda v = g$ for an arbitrary constant λ if we set $v = e^{-\lambda t} u$ and $g = e^{\lambda t} f$. Finally, in case of the boundary point for the heat equation discussed above we show in [1] that at these points the solution u is not C^∞, even if all the compatibility conditions are ratisfied, its smoothness depends on the size of M (in (31)) which controls the size of γ. Furthermore, we should remark that this theorem is global in nature, in other words the smoothness of f at one point has influence on the smoothness of u at every point. In the rest of this lecture we will discuss some general global results concerning the smoothness of solutions in the variational problems introduced in the first lecture. We will deal with the case when Q is of first order, i.e.

(43) $$Q(\varphi, \psi) = \sum_{|\alpha|, |B| \leq 1} (a^{ij}_{\alpha\beta} D^\alpha \varphi_i, D^\beta \psi_j) ;$$

the higher order case can be reduced to this (see also [4] for a direct treatment of the higher order case).

Let $\Omega \subset \mathbb{R}^n$ be a bounded domain with a smooth boundary $b\Omega$ defined by $r = 0$, where $r \in C^\infty$ and $\mathrm{grad}(r) \neq 0$ near $b\Omega$. If $P \in b\Omega$, there is a neighborhood U of P and functions $t_1, \ldots, t_{n-1} \in C^\infty(U)$ such that $t_1, \ldots, t_{n-1}, r$ form a coordinate system on U with origin P. Now if $\mathcal{D}$ is the domain of Q we set $\dot{\mathcal{D}} = \mathcal{D} \cap (C^\infty(\bar{\Omega}))^m$ and assume

(a) $\dot{\mathcal{D}}$ is dense in $(L_2(\Omega))^m$ and $(C_0^\infty(\Omega))^m \subset \dot{\mathcal{D}}$.

(b) $\mathcal{D}$ is complete under Q.

(c) If $f \in C^\infty(\bar{\Omega})$ and $\varphi \in \dot{\mathcal{D}}$ then $f\varphi \in \dot{\mathcal{D}}$.

(d) $\zeta \in C_0(U)$, $\varphi \in \dot{\mathcal{D}}$ with $\varphi = (\varphi_1, \ldots, \varphi_m)$ then

$$\left(\frac{\partial}{\partial t_i}(\zeta\varphi), \ldots, \frac{\partial}{\partial t_i}(\zeta\varphi)\right) \in \mathcal{D} \quad \text{for } i = 1, \ldots, n-1.$$

(e) Suppose $V \subset \bar{V} \subset U$ and $h = (h_1, \ldots, h_{n-1})$ so that $(t+h, r) \in U$ whenever $(t, r) \in V$ then we assume that if $\varphi \in \dot{\mathcal{D}}$ with $\mathrm{supp}(\varphi) \subset V \cap \bar{\Omega}$ then $T_h\varphi \in \mathcal{D}$, where $T_h(t, r) = \varphi(t+h, r)$. Furthermore if $K : C_0^\infty(U \cap \bar{\Omega}) \to C_0^\infty(U \cap \Omega)$ such that $KT_h u = T_h Ku$ for $|h|$ sufficiently small then $K\varphi \in \dot{\mathcal{D}}$.

44. <u>Definition.</u> Q is <u>non-characteristic</u> at $P \in b\Omega$ if, relative to the coordinates introduced above $a^{ij}_{hn}(P)$ is non-singular.

45. <u>Theorem.</u> Under the assumption of Theorem 6, if Q is non-characteristic for each $P \in b\Omega$, if $\mathcal{D}$ satisfies the conditions (a) to (e) and if Q is

compact relative to L_2 (i.e. a sequence in $\mathcal{D}$ bounded with respect to the Q-norm has a convergent subsequence in L_2). Then if $\alpha \in (C^\infty(\bar{\Omega}))^m$ there exists a unique $\varphi \in \mathcal{D}$ such that $Q(\varphi,\psi) = (\alpha,\psi)$ for all $\psi \in \mathcal{D}$ and then $\varphi \in \dot{\mathcal{D}}$. Furthermore, the operator that takes α to φ is completely continuous and its eigenspaces are in $\dot{\mathcal{D}}$.

46. Definition. If $u \in C^\infty(\Omega)$ we define $\|u\|_m$ for a non-negative integer m by

$$\|u\|_m^2 = \sum_{|\alpha| \leq m} \|D^\alpha u\|^2, \tag{47}$$

if $m < 0$ we define $\|u\|_m$ by

$$\|u\|_m = \max_{v \in C^\infty(\bar{\Omega})} \cdot \frac{(u,v)}{\|v\|_{-m}} . \tag{48}$$

We denote by H_m the completion of $C^\infty(\bar{\Omega})$ under the norm $\| \ \|_m$.

As in the discussion of the last equation, to prove the smoothing of the solution φ if we proceed by first establishing a priori estimates for the derivatives. This is done with the help of the following.

49. Lemma. Suppose that Q satisfies (7), (8) and (9). Then Q is compact with respect to L_2 if and only if for each $\varepsilon > 0$ there exists a constant $C(\varepsilon) > 0$ such that:

$$\|\varphi\|^2 \leq \varepsilon Q(\varphi,\varphi) + C(\varepsilon)\|\varphi\|_{-1}^2 \tag{50}$$

for all $\varphi \in \mathcal{D}$. Here we set $\|\varphi\|_m^2 = \sum_j \|\varphi_j\|_m^2$.

The above lemma follows by a standard argument from the fact that the L_2-norm is compact with respect to the -1 norm.

To estimate the mth derivatives of φ we proceed as follows.

We first choose a covering of $\bar{\Omega}$ by neighborhoods $U_0, U_1, \ldots, U_N$ such that U_ν, if $v \geq 1$, admits a boundary coordinates system (t, r) and $\bar{U}_0 \cap b\Omega = \phi$. Let $\{\zeta_0, \ldots, \zeta_N\}$ be a partition of unity on $\bar{\Omega}$ subordinate to this covering. Now on a fixed U_ν with $v \geq 1$, let $\beta = (\beta_1, \ldots, \beta_{n-1})$ with $\Sigma\, \beta_j \leq m$. We apply (50) to $D_t^\beta(\zeta_\nu\varphi) = (D_t^\beta \zeta_\nu \varphi_1, \ldots, D_t^\beta \zeta_\nu \varphi_k)$.

51. Lemma. Under the above assumptions:

$$\sum_{|\beta| \leq m} |Q(D_t^\beta \zeta_\nu \varphi, D_t^\beta \zeta_\nu \varphi) - Q(\varphi, \zeta_\nu D_t^{2\beta} \zeta_\nu \varphi)|$$
$$\leq C\Big(\sum_{|\gamma| \leq m-1} \|D_t^\gamma D_r \zeta_\nu' \varphi\|^2 + \sum_{|\gamma| \leq m} \|D_t \zeta_\nu' \varphi\|^2 \Big),$$

where $\zeta_\nu' \in C_0^\infty(U_\nu \cap \bar{\Omega})$, $\zeta_\nu' = 1$ on the support of ζ_ν and the constant C depends on m.

If φ is the solution to the variational problem in Theorem 45 then

$$Q(\varphi, \zeta_\nu D_t^{2\beta} \zeta_\nu \varphi) = (D_t^\beta \zeta_\nu \alpha, D_t^\beta \zeta_\nu \varphi) \leq C \|\alpha\|_m \|\varphi\|_m . \tag{52}$$

Since Q is non-characteristic we have:

$$\|D_r \zeta_\nu \varphi\|^2 \leq \text{const}\Big(Q(\zeta_\nu \varphi, \zeta_\nu \varphi) + \sum_{|\beta| \leq 1} \|D_t^\beta \zeta_\nu \varphi\|^2\Big) \tag{53}$$
$$\leq \text{const}\Big(\|\alpha\|^2 + \sum_{|\beta| \leq 1} \|D_t^\beta \zeta_\nu \varphi\|^2\Big)$$

Furthermore φ satisfies the equations:

(54) $$\sum_{|\alpha|, |\beta|=1} a^{ij}_{\alpha\beta} D^{\alpha+\beta} \varphi_i + \text{lower order terms} = \alpha_j .$$

Hence, by the non-characteristicity, we have

(55) $$D_r^2 \varphi_i = \sum_{|\gamma|\leq 2} b^j_{i\gamma} D_t^\gamma \varphi_j + \sum_{|\gamma|\leq 1} c^j_{i\gamma} D_t^\gamma D_r \varphi_j .$$

Replacing $\zeta_\nu \varphi$ in (53) by $D_t^\gamma \zeta_\nu \varphi$ and using Lemma 51 and (52) we obtain

(56) $$\sum_{|\gamma|\leq m-1} \| D_r D_t^\gamma \zeta_\nu \varphi \|^2 \leq C(\|\alpha\|^2_{m-1} + \sum_{|\beta|\leq m} \| D_t^\beta \zeta'_\nu \varphi \|^2) .$$

Differentiating (55) repeatedly and multiplying by ζ_ν taking L_2-norms and combining with the above we obtain the appriori estimate,

(57) $$\|\zeta_\nu \varphi\|^2_m \leq \varepsilon \text{ const.} (\|\alpha\|^2_m + \|\zeta'_\nu \varphi\|^2_m) + C(\varepsilon)\text{const.} \|\varphi\|^2_{m-1}$$

The same estimate holds for $v = 0$, the proof being simpler since we do not have normal derivatives. Summing over γ choosing ε small we obtain:

(58) $$\|\varphi\|_m \leq C_m(\|\alpha\|_m + \|\varphi\|_{m-1})$$

and since $\|\varphi\| \leq \text{const.} \|\alpha\|$ we obtain by induction on m:

(59) $$\|\varphi\|_m \leq \text{const.} \|\alpha\|_m .$$

Now we have to show that the derivatives of φ are actually square integrable. It would be hard to apply the smoothing operator directly as in the first lecture. Instead we use the method of "elliptic regularization".

Suppose $\widetilde{Q}$ is a form defined over $\mathcal{D}$ which satisfies

(60) $$\|\varphi\|_1^2 \leq \text{const. } \widetilde{Q}(\varphi, \varphi)$$

for all $\varphi \in \dot{\mathcal{D}}$. Then by standard elliptic theory it is shown that if $\widetilde{Q}(\varphi,\psi) = (\alpha, \psi)$ for all $\psi \in \mathcal{D}$ then φ is smooth whenever α is smooth (this can in fact be shown by the same techniques as in the first lecture).

Now we set Q^δ to be the form on $\dot{\mathcal{D}}$ defined by:

(61) $$Q^\delta(\varphi, \psi) = Q(\varphi, \psi) + \delta \sum_i \sum_{|\gamma|=1} (D^\gamma \varphi_i, D^\gamma \psi_i)$$

with $\delta \geq 0$.

Then, given $\alpha \in (C^\infty(\overline{\Omega}))^m$ there exists a unique $\varphi^\delta \in \dot{\mathcal{D}}$ such that

(62) $$Q^\delta(\varphi^\delta, \psi) = (\alpha, \psi) \text{ for all } \psi \in \dot{\mathcal{D}} .$$

Furthermore, examining the proof of (59) we see that the estimate

(63) $$\|\varphi^\delta\|_m \leq \text{const. } \|\alpha\|_m$$

holds with a constant that is independent of δ (for δ small and $\delta \geq 0$). Hence as $\delta \to 0$ the φ^δ have a subsequence that goes to a limit $\varphi^0 \in C^\infty(\overline{\Omega})$ but now for $\psi \in \mathcal{D}$ we have

(64) $$Q(\varphi^0, \psi) = \lim Q(\varphi^\delta, \psi) = \lim(Q^\delta(\varphi^\delta, \psi) - \delta \sum_{|\gamma|=1} (D^\gamma \varphi^\gamma, D^\gamma \psi))$$
$$= (\alpha, \psi) .$$

Therefore $\varphi = \varphi^0$, by uniqueness.

From compactness it follows that there is a discrete set of $\{\lambda\}$ such that

$$Q(\varphi, \psi) = \lambda(\varphi, \psi) \tag{65}$$

for some $\varphi \in \mathcal{D}$, $\varphi \neq 0$ and all $\psi \in \mathcal{D}$. Let S_λ denote the space of all such φ and let $\Pi_\lambda : \mathcal{D} \to S_\lambda$ denote orthogonal projection. For fixed λ, $\varepsilon \neq 0$ and $\alpha \in L_2(\Omega)$ there exists a unique φ^ε such that

$$Q(\varphi, \psi) - \lambda(\varphi, \psi) - \varepsilon(\Pi_\lambda \varphi, \psi) = (\alpha, \psi), \tag{66}$$

To see this ,note that the left hand side cannot be zero for all ψ unless $\varphi = 0$; if it were set $\psi = \Pi_\lambda \varphi$ we conclude $\Pi_\lambda \varphi = 0$ so that $\varphi \in S_\lambda$ but then $\varphi = \Pi_\lambda \varphi = 0$. Hence by Fredholm theory a unique $\varphi \in \mathcal{D}$ exists. If $\alpha \in H_m$ (i.e. $\|\alpha\|_m < \infty$) then arguing as above we can show that $\varphi \in H_m$ and

$$\|\varphi\|_m \leq C_m(\|\alpha\|_m + \|\varphi\|) . \tag{67}$$

We wish to show that $S_\lambda \subset \mathcal{D}$ let $\sigma_1, \ldots, \sigma_q$ be an orthonormal basis of S and assume that $\sigma_1, \ldots, \sigma_k \in \dot{\mathcal{D}}$, with $k < q$. We will construct an element $\theta \in \dot{\mathcal{D}}$ such that $(\theta, \sigma_j) = 0$ for $j = 1, \ldots, k$ and $\|\theta\| = 1$. Let $\alpha \in C^\infty(\overline{\Omega})$ such that $(\alpha, \sigma_j) = 0$ for $j = 1, \ldots, k$ and $(\alpha, \sigma_{p+1}) \neq 0$. Let φ^ε be the solution of (66), then $\varphi^\varepsilon \in \dot{\mathcal{D}}$. We can find a sequence $\varepsilon_j \to 0$ such that $\|\varphi^{\varepsilon_j}\| \to \infty$ for if such a sequence did not exist then by (67) $\|\varphi^{\varepsilon_j}\|_m$ would be bounded for all m and hence would have a covergent subsequence whose limit φ^0 would satisfy the equation

(68) $$Q(\varphi^0, \psi) - \lambda(\varphi^0, \psi) = (\alpha, \psi) ,$$

which is impossible since α is not orthogonal to S_λ.

Let $\theta_\nu = \dfrac{\varphi^{\varepsilon_\nu}}{\|\varphi^{\varepsilon_\nu}\|}$ then by (67) we can find a subsequence which converges in H_m, for every m, so the limit θ is C^∞ and hence $\theta \in \dot{\mathcal{D}}$. Further, by (66) we have

$$Q(\theta, \psi) - \lambda(\theta, \psi) = 0$$

hence $\theta \in S_\lambda$. Now setting $\psi = \sigma_j$ in (66) with $j \leq k$ we obtain

$$(\Pi_\lambda \theta_\nu, \sigma_j) = 0$$

and hence $(\theta, \sigma_j) = 0$ for $j = 1, \ldots, k$ which concludes the proof of Theorem 45.

Lecture 3

In this lecture we wish to discuss local regularity. Consider, for example a bounded domain $\Omega \subset \mathbb{R}^2$. Given $f \in C^\infty(\overline{\Omega})$ it is clear that there exists a unique solution u of $\frac{\partial u}{\partial x} = f$ with $u = 0$ on $b\Omega$, furthermore $u \in C^\infty(\overline{\Omega})$. This problem, however, does not have local regularity; suppose that part of the boundary is given by, $y = 0$ and let $f(x, y) = \rho(x)\sigma(y)$ in a neighborhood of this part of $b\Omega$. Suppose further that $\rho(0) \doteq 0$ that $\rho \geq 0$, that $\rho(x) > 0$ for $0 < x < a$ and $\rho(x) = 0$ for $a \leq x \leq b$. Then if σ is not differentiable it is clear that the solution u will not be differentiable in a rectangle which has (a, b) even though $f = 0$ on this rectangle.

Consider the operator

$$(67) \qquad Lu = \sum_{j=1}^{k} X_j^2 u + X_0 u + cu$$

where the X_j are real C^∞ vector fields defined in an open subset U of $\mathbb{R}^n$. In [5] Hörmander proves the following.

68. Theorem. If the Lie algebra generated by $X_0, X_1, \ldots, X_k$ over C^∞ functions, when evaluated at P, contains all tangent vectors at P and if this condition is satisfied for each $P \in U$ then L is hypoelliptic.

Here we will outline a proof of this theorem along the lines of [6] organized in such a way that it will introduce the methods described in [14]. For fixed $P \in U$ we want to show that, under the hypotheses of the theorem, if f is C^∞ in a neighborhood of P then every solution is of $Lu = f$ is C^∞ in a neighborhood of P. For this it suffices to show that there exists a neighborhood V of P and positive constants $\varepsilon > 0$, $c > 0$ such that

$$(69) \qquad \|u\|_\varepsilon \leq c(\|Lu\| + \|u\|) \quad \text{for all } u \in C_0^\infty(V),$$

where $\|u\|_\varepsilon$ denotes the usual Sobolev norm. First we want to point out that if L were an arbitrary second order operator the inequality (69) would not imply hypoellipticity, in fact, the wave operator ($Lu = u_{xx} - u_{yy}$) satisfies this inequality with $\varepsilon = 1$.

70. Lemma. If L is given by (67) and if (69) holds then L is hypoelliptic. The main point is the following inequality, which is easily proven by

integration by parts:

(71) $$\sum_{j=1}^{k} \|X_j u\|^2 \leq |(Lu, u)| + c\|u\|^2 ,$$

this holds for all $u \in C_0^\infty(V)$ with some fixed $c > 0$, where $\overline{V} \subset U$.

Let P be a pseudo-differential operator of order s, the following gives a control of $[L, P]$.

72. Lemma. If P is a pseudo-differential operator of order s then there exists a constant $c > 0$ such that

(73) $$\|[L, P]u\| \leq c(\|Lu\|_s + \|u\|_s)$$

for all $u \in C_0^\infty(V)$.

Proof: First we show that

(74) $$\sum_{j=1}^{k} \|X_j Pu\|^2 \leq |(LPu, Pu)| + \text{const.} \|Pu\|^2 ,$$

for all $u \in C_0^\infty(V)$. Choose $\zeta \in C_0^\infty(U)$ with $\zeta = 1$ on V. Let V' be a neighborhood of $\sup(\zeta)$ with $\overline{V}' \subset U$; then (71) is also valid (with a different c) from functions in $C_0^\infty(V')$. Thus we can apply (71) to ζPu, we observe that

(75) $$\zeta Pu = P\zeta u + [\zeta, P]u = Pu + [\zeta, P]u$$

for all $u \in C_0^\infty(V)$. Furthermore since the support of u is disjoint from the support of the total symbol of $[\zeta, P]$ we have, for every m.

(76) $$\|[\zeta, P]u\|_m \le C_m \|u\|, \quad \text{for all } u \in C_0^\infty(V) .$$

Thus (74) follows. Next observe that

(77) $$[L, P] = 2 \sum_{j=1}^{k} [X_j, P]X_j + \Sigma [X_j, [X_j, P] + [X_0, P] - [P, \ ]$$

$$= \sum_{j=1}^{k} T_j X_j + S = \sum_{1}^{k} X_j T_j + S'$$

where the T_j, S and S' are pseudo-differential operators of degree s.

Thus we see that

(78) $$\|[L, P]u\|^2 \le \text{const.} \left(\sum_{1}^{k} \|X_j u\|_s^2 + \|u\|_s^2\right) ,$$

for all $u \in C_0^\infty(V)$. We denote by Λ^s the pseudo-differential operator with symbol $(1+|\xi|^2)^{s/2}$. Then

(79) $$\sum_{1}^{k} \|X_j u\|_s^2 = \sum_{1}^{k} \|\Lambda^s X_j u\|^2 \le \sum_{1}^{k} \|X_j \Lambda^s u\|^2 + \text{const. } \|u\|_s^2$$

$$\le |(L\Lambda^s u, \Lambda^s u)| + \text{const. } \|u\|_s^2$$

$$\le \|Lu\|_s^2 + |([L, \Lambda^s]u, \Lambda^s u)| + \text{const. } \|u\|_s^2$$

$$\le \|Lu\|_s^2 + \text{const. } \left(\sqrt{\sum_{1}^{k} \|X_j u\|_s^2}\, \|u\|_s + \|u\|_s^2\right)$$

it then follows that

(80) $$\sum_{1}^{k} \|X_j u\|_s^2 \le \text{const. } (\|Lu\|_s^2 + \|u\|_s^2) \quad \text{for all } u \in C_0^\infty(V) .$$

and (73) follows by substituting this into (78).

Arguing as above we can replace u by $\Lambda^{k\varepsilon}u$ in (69), (this, of course, requires increasing C appropriately). Then applying (73) we obtain

$$\|u\|_{(k+1)\varepsilon} \leq \text{const.}\ (\|Lu\|_{k\varepsilon} + \|u\|_{k\varepsilon}) \tag{81}$$

and hence

$$\|u\|_{s+\varepsilon} \leq \text{const.}\ (\|Lu\|_{s} + \|u\|) \tag{82}$$

for all $u \in C_0^\infty(V)$.

Now suppose $u \in C^\infty(U)$ and $\zeta \in C_0^\infty(V)$ we then replace u by ζu. Let $\zeta' \in C_0^\infty(V)$ such that $\zeta' = 1$ on the support of ζ, then $\zeta u = \zeta\zeta' u$ and we have replacing P by ζ in (77) and observing that then $\operatorname{supp}(T_j) \subset \operatorname{supp}(\zeta)$ and $\operatorname{supp}(S) \subset \operatorname{supp}(\zeta)$. Thus, instead of (81) we obtain:

$$\|\zeta u\|_{(k+1)\varepsilon} \leq \text{const.}\ (\|\zeta Lu\|_{k\varepsilon} + \|\zeta' u\|_{k\varepsilon} + \|u\|) \tag{83}$$

by induction on k (and choosing an increasing sequence of $\zeta' \in C_0^\infty(V)$, each of which equals one or the support of the previous one) we obtain:

$$\|\zeta u\|_{(k+1)\varepsilon} \leq \text{const.}\ (\|\zeta' Lu\|_{k\varepsilon} + \|\zeta' u\|). \tag{84}$$

To obtain hypoellipticity from these estimates we proceed either by using a smoothing operator, as in the first lecture (replacing P in (77) by the smoothing operator) or by using elliptic regularization as in the second lecture (deriving the estimates of $L_\delta = L + \delta\Delta$ with constants independent of δ). This proves Lemma 70.

To prove Theorem 78 we will prove (69) under the hypotheses of the theorem. First let $Y_{i_1,\ldots,i_p}$ be defined by Y_{i_1} and $Y_{i_1\ldots i_p} = [X_{i_p}, Y_{i_1\ldots i_{p-1}}]$. From the Jacobi identity and the hypothesis it follows that there are a finite number of the Y's which span all vector fields in a neighborhood of $\bar{V}$. Now let $\mathcal{P}$ be the set of all pseudo-differential operators of order zero such that if $P \in \mathcal{P}$, then there exists $\varepsilon > 0$ and $c > 0$ such that:

$$\text{(85)} \qquad \|Pu\|_\varepsilon \leq c(\|Lu\| + \|u\|)$$

for all $u \in C_0^\infty(V)$.

The set satisfies the following properties:

(A) $\mathcal{P}$ is ideal (i.e. if $P \in \mathcal{P}$ and Q is a pseudo differential operator of order 0 then $PQ \in \mathcal{P}$ and $QP \in \mathcal{P}$. Further if $P \in \mathcal{P}$ then $P^* \in \mathcal{P}$.

(B) $X_j \Lambda^{-1} \in \mathcal{P}$ for $j = 0, \ldots, m$.

(C) If $P \in \mathcal{P}$ then $[X_j, P] \in \mathcal{P}$ for $j = 0, \ldots, m$.

It then follows inductively from (C) that $Y_{i_1\ldots i_p} \Lambda^{-1} \in \mathcal{P}$ and hence by every 0 order pseudo-differential operator in $\mathcal{P}$ which would prove the theorem. It remains to establish the above properties. (A) is immediate. For $j > 0$ we have

$$\text{(86)} \qquad \|X_j \Lambda^{-1} u\|_1^2 \leq \|X_j u\|^2 + \text{const.}\ \|u\|^2$$

so (71) implies that $X_j \Lambda^{-1} \in \mathcal{P}$ for $j > 0$. Now consider

(87) $$\|X_0\Lambda^{-1}u\|^2_{\frac{1}{2}} \leq (X_0u, Tu) + \text{const. } \|u\|^2$$

$$\leq (Lu, Tu) - \sum_{j=1}^{k} (X_j^2u, Tu) + \text{const. } \|u\| ,$$

where T is a zero-order pseudo-differential operator. Then

(88) $$|(X_j^2u, Tu)| \leq \text{const. } (\|X_ju\|^2 + \|u\|^2)$$

and hence we obtain (again using (71)) that $X_0\Lambda^{-1} \epsilon \mathcal{P}$.

To establish (c) suppose $P \epsilon \mathcal{P}$ so (85) holds, now consider:

(89) $$\|[X_j, P]u\|^2_\delta = (X_jPu, T^{2\delta}u) - (PX_ju, T^{2\delta}u),$$

when $T^{2\delta}$ is a pseudo-differential operator of order 2δ. It then follows that for $j > 0$

(90) $$\|[X_j, P]u\|^2_\delta \leq \text{const. } (\|X_ju\|^2 + \|Pu\|^2_{2\delta} + \|u\|^2_{2\delta - 1})$$

so that taking $2\delta \leq \min(1, \varepsilon)$ we obtain $[X_j, P] \epsilon \mathcal{P}$ for $j > 0$. Consider (89) with $j = 0$, then

(91) $$|(X_0Pu, T^{2\delta}u)| \leq |(Pu, T^{2\delta}X_0u)| + c(\|Pu\|^2_{2\delta} + \|u\|)$$

$$\leq |(Pu, T^{2\delta}Lu)| + |\sum_1^k (Pu, T^{2\delta}X_j^2u)| + C(\|Pu\|^2_{2\delta} + \|u\|)$$

$$\leq \|Lu\|^2 + \text{const.} (\sum_1^k \|X_jPu\|^2_{2\delta} + \|Pu\|^2_{2\delta} + \|u\|^2)$$

From (79) and (77) it follows that

(92) $$\sum_1^k \|X_ju\|^2_s \leq \text{const.} (\|Lu\|^2 + \|u\|^2_{2s}) .$$

Combining this and (73) with (91) we obtain

(93) $$|(X_0 Pu, T^{2\delta} u)| \leq \text{const.}(\|Lu\|^2 + \|Pu\|^2_{4\delta} + \|u\|^2).$$

Similarly we estimate last term of (89) by

(94) $$|(X_0 Pu, T^{2\delta} u)| \leq \text{const.}(\|Lu\|^2 + \|Pu\|^2_{4\delta} + \|u\|^2).$$

Hence we conclude that $X_0 \Lambda^{-1} \in \mathcal{P}$ which establishes the theorem.

For the most precise estimates of these operators we refer to [8], where an approximate parametrix for L is constructed.

Lecture 4.

We will now consider the more general situation of a first order quadratic form Q defined on $\mathcal{D}$ by (43) and we assume that $\mathcal{D}$ satisfies properties (a) to (c) of the second lecture. Suppose $P \in b\Omega$ we say that Q is subelliptic at P if there exists a neighborhood U of P and positive contents ε and C such that

(95) $$\|\varphi\|^2_{\varepsilon} \leq CQ(\varphi, \varphi) \quad \text{for all } \varphi \in \mathcal{D}_{U},$$

where $\dot{\mathcal{D}}_U$ consists of all $\varphi \in \dot{\mathcal{D}}$ with $\text{supp}(\varphi) \subset U \cap \bar{\Omega}$.

96. Theorem. If Q satisfies the conditions of Theorem 6, $\mathcal{D}$ satisfies properties (a) to (e) and if Q is subelliptic and non-characteristic at P, then the following local regularity holds. If $\alpha \in L_2(\Omega)$ and α is C^∞ on $U \cap \bar{\Omega}$ then the unique solution φ of $Q(\varphi, \psi) = (\alpha, \psi)$ is also C^∞ on $U \cap \bar{\Omega}$.

In fact if $\zeta, \zeta' \in C_0^\infty(U)$ and if $\zeta' = 1$ on the support of ζ then

(97) $$\|\zeta\varphi\|_{s+2\varepsilon} \leq \text{const. } \|\zeta'\alpha\|_s .$$

Furthermore, the solution φ of $Q(\varphi,\psi) = \lambda(\varphi, \psi)$ are also C^∞ in $U \cap \bar{\Omega}$.

The proof of this proposition is derived by first obtaining a-priori estimates for derivatives of φ. Let $\zeta, \zeta' \in C_0^\infty(U)$ with $\zeta' = 1$ on the support of ζ then substituting. Suppose (t, r) boundary coordinates on U. We define for $u \in C_0^\infty(U \cap \bar{\Omega})$ the partial Fourier transform $\tilde{u}(\tau, r)$ by:

(98) $$\tilde{u}(\tau, r) = \int_{\mathbb{R}^{n-1}} e^{-it\cdot\tau} \mu(t, r)dt;$$

where $t \cdot \tau = \sum_1^{n-1} t_i \tau_i$ and $dt = dt_1 \ldots dt_{n-1}$. Let Λ_t^s denote the tangential pseudo-differential operator defined by:

(99) $$\widetilde{\Lambda_t^s u}(\tau, r) = (1+|\tau|^2)^{s/2}\, \tilde{u}(\tau, r) .$$

Then if $\varphi \in \dot{\mathcal{D}}$, we have $\zeta'\Lambda_t^s\zeta\varphi \in \dot{\mathcal{D}}$. Replacing φ in (95) by this we have:

(100) $$\|\zeta'\Lambda_t^s\zeta\varphi\|_\varepsilon^2 \leq CQ(\zeta'\Lambda_t^s\zeta\varphi, \zeta'\Lambda_t^s\zeta\varphi)$$

for all $\varphi \in \dot{\mathcal{D}}_0$.

The following is proved in [1] (Lemma 3.1).

101. Lemma. If $\zeta'' \in (C_0^\infty(U)$ and $\zeta'' = 1$ on the support of ζ then

$$Q(\zeta'\Lambda_t^s\zeta\varphi, \zeta'\Lambda_t^s\zeta\varphi) = Q(\varphi, \zeta\Lambda_t^s(\zeta')^2\Lambda_t^s\zeta\varphi)$$

$$+ 0(\|\Lambda_t^{s-1}\zeta''\varphi\|_1^2) .$$

The non-characteristicity of Q implies that

$$\|\zeta'\Lambda_t^{s-1}\zeta\varphi\|_1^2 \leq C(Q(\zeta'\Lambda_t^{s-1}\zeta\varphi, \zeta'\Lambda_t^{s-1}\zeta\varphi) + \|\Lambda_t^s\zeta\varphi\|^2) . \tag{102}$$

The theorem is then obtained by following the same scheme as in the proof of Theorem 45.

The following result gives an important class of Q that are subelliptic with $\varepsilon = \frac{1}{2}$.

103. Theorem. If Q is elliptic on $(C_0^\infty(\Omega))^k$ i.e. there exists $C > 0$ suc that

$$\|\varphi\|_1^2 \leq CQ(\varphi,\varphi) , \tag{104}$$

for all $\varphi \in (C_0^\infty(\Omega))^k$ and if there exists a constant $C' > 0$ such that:

$$\int_{b\Omega} |\varphi|^2 ds \leq C' Q(\varphi,\varphi) \text{ for all } \varphi \in \dot{\mathcal{D}} , \tag{105}$$

where ds denotes the volume element on $b\Omega$. Then there exists $C'' > 0$ such that

$$\|\varphi\|_{\frac{1}{2}}^2 \leq C'' Q(\varphi, \varphi) . \tag{106}$$

This result is proved in [1], (Theorem 5) the idea of the proof is to apply (104) to $\Lambda_t^{-\frac{1}{2}}(\varphi, \varphi^0)$ where $\sup(\varphi) \subset U \cap \bar{\Omega}$ and where φ^0 is defined by:

(107) $$\tilde{\varphi}^0(\tau, r) = e^{(1+|\tau|^2)^{\frac{1}{2}} r}\tilde{\varphi}(\tau, 0),$$

then

(108) $$\|\Lambda_t^{-\frac{1}{2}}\varphi^0\|_1^2 \leq 2\int_{b\Omega} |\varphi|^2\, ds.$$

The result follows by using standard arguments in elliptic theory.

The type of variational problem that we have been discussing arises from the study of systems of equations as follows. Suppose $A : (C^\infty(\bar{\Omega}))^p \to (C^\infty(\bar{\Omega}))^k$ is a first order operator given by

(109) $$(Au)_i = \sum_j \sum_{|\alpha| \leq 1} a_{i\alpha}^j D^\alpha u_j \qquad i=1, \ldots, k$$

where $a_{i\alpha}^j \in C^\infty(\bar{\Omega})$. Given $\alpha \in (C^\infty(\bar{\Omega}))^k$ (or in $(L_2(\Omega))^k$) we wish to solve the equation:

(110) $$Au = \alpha.$$

Let $B : (C^\infty(\bar{\Omega}))^k \to (C^\infty(\bar{\Omega}))^q$ given by

(111) $$(B\varphi)_s = \sum_{t=1}^k \sum_{|\beta| \leq 1} b_{s\beta}^t D^\beta \varphi_t, \qquad s=1, \ldots, q$$

such that

(112) $$BA = 0.$$

Then, clearly, a necessary condition for the solvability of (110) is

(113) $$B\alpha = 0 .$$

Let $\psi \in \mathrm{Dom}(A^*)$ with $A^*\psi = 0$ then

(114) $$(Au, \psi) = (u, A^*\psi) = 0 .$$

Hence, another necessary condition for the solvability of (110) is that α is orthogonal to the null space of A^* (which we denote by $\mathcal{N}(A^*)$).

115. Theorem. Let $\mathcal{D} = \mathrm{Dom}(A^*) \cap \mathrm{Dom}(B)$. Suppose that there exists $c > 0$ such that

(116) $$\|\varphi\|^2 \leq c(\|A^*\varphi\|^2 + \|B\varphi\|^2) \text{ for all } \varphi \in \mathcal{D}.$$

Then if $\alpha \in (L_2(\Omega))^k$ with $B\alpha = 0$ and $\alpha \perp \eta(A^*)$ there exists a $u \in \mathrm{Dom}(A$ satisfying (110).

Outline of proof: $\mathcal{H} = \{\varphi \in \mathcal{D} \mid A^*\varphi = 0,\ B\varphi = 0\}$. Let the "Laplacian" L be defined by

(117) $$L = AA^* + B^*B$$

with $\mathrm{Dom}(L) = \{\varphi \in \mathcal{D} \mid B\varphi \;\; \mathrm{Dom}(A^*)$ and $A^*\varphi \in \mathrm{Dom}(A)\}$. Then $\eta(L) = \mathcal{H}$ since clearly $\mathcal{H} \subset \eta(L)$ and if $\varphi \in \mathrm{Hom}(L)$ then

(118) $$(L\varphi,\varphi) = \|A^*\varphi\|^2 + \|B\varphi\|^2$$

so if $\varphi \in \eta(L)$ then $\varphi \in \mathcal{H}$.

Observe that L is self-adjoint and that (116) together with (118) imply that the range of L is closed. Therefore, the necessary and sufficient condition for solving

(119) $$L\varphi = \alpha$$

is that α is orthogonal to $\mathcal{H}$. Now suppose $B\alpha = 0$ and α is orthogonal to $\eta(A^*)$, then α is orthogonal to $\mathcal{H}$ and hence we can solve (119) so, we have

(120) $$AA^*\varphi + B^*B\varphi = \alpha ,$$

applying B to this equation we obtain

(121) $$BB^*B\varphi = 0 ,$$

taking inner products with $B\varphi$ we have

(122) $$(BB^*B\varphi, B\varphi) = \|B^*B\varphi\|^2 = 0 ,$$

so that (120) implies that:

(123) $$A(A^*\varphi) = \alpha ,$$

hence $u = A^*\varphi$ is a solution of (110). Observe that this is the unique solution which is orthogonal to $\eta(A)$.

Let $\mathcal{D} = \mathrm{Dom}(A^*) \cap \mathrm{Dom}(B)$ in the sense of L_2, now on $\mathcal{D}$ we define the form Q by:

(124) $$Q(\varphi,\psi) = (A^*\varphi, A^*\psi) + (B\varphi, B\psi) + (\varphi,\psi)$$

Observe that $\dot{\mathcal{D}} = \mathrm{Dom}(A^*) \cap (C^\infty(\bar{\Omega}))^k$, now $\varphi \in \dot{\mathcal{D}}$ if and only if $\varphi \in (C^\infty(\bar{\Omega}))^k$ and

(125) $$(Au, \varphi) = (u, {}^t\bar{A}\varphi) \text{ for all } u \in (C^\infty(\bar{\Omega}))^p$$

but for any $\varphi \in (C^\infty(\bar{\Omega}))^k$ we have

$$(Au, \varphi) = (u, {}^t\bar{A}\varphi) + \int_{b\Omega} \sigma({}^t\bar{A}, dr)\varphi\bar{u}\, ds,$$

where $\sigma({}^t\bar{A}, dr)$ denotes the symbol of ${}^t\bar{A}$ evaluated at dr. Have we concluded that $\varphi \in \dot{\mathcal{D}}$ if and only if:

(126) $$\sigma({}^t\bar{A}, dr)\varphi = 0 \text{ on } b\Omega .$$

127. Theorem. If Q is given by (124) is non-characteristic, if the matrix $\sigma({}^t\bar{A}, dr)$ has constant rank on $b\Omega$ and if Q is compact with respect to the L_2-norm, then for $\alpha \in (L_2(\Omega))^k$ satisfying the conditions $B\alpha = 0$ and $\alpha \perp \mathcal{N}(A^*)$ there exists a solution to (110). If $\alpha \in (C^\infty(\bar{\Omega}))^k$ then the unique solution is orthogonal to (A) is in $(C^\infty(\bar{\Omega}))^p$. Furthermore, the eigen-spaces of Q are finite dimensional and contained in $\dot{\mathcal{D}}$.

Under the hypotheses of the above theorem consider the operator; given by (117); there exists then a completely continuous operator N on $(L_2(\Omega))^k$ defined as follows. If $\alpha \perp \mathcal{H}$ then $N\alpha$ is the unique solution of $LN\alpha = 0$ which is orthogonal to $\mathcal{H}$, if $\alpha \perp \mathcal{H}$ then $N\alpha = 0$. Observe that if Q is subelliptic then N is pseudo-local and that if Q is compact then N is completely continuous.

Lecture 5

In this lecture we will introduce the $\bar{\partial}$-Neumann problem (see [9],

[10]). Suppose $\Omega \subset \mathbb{C}^n$ is a bounded domain with a smooth boundary. Let $z_1, \ldots, z_n$ denote the coordinates on $\mathbb{C}^n$ and $x_j = \mathrm{Re}(z_j)$, $y_j = \mathrm{Im}(z_j)$; then, as usual, we define the operators:

(128) $$\frac{\partial}{\partial z_j} = \frac{1}{2}\left(\frac{\partial}{\partial x_j} - i\frac{\partial}{\partial y_j}\right) \text{ and } \frac{\partial}{\partial \bar{z}_j} = \frac{1}{2}\left(\frac{\partial}{\partial x_j} + i\frac{\partial}{\partial y_j}\right).$$

Consider the system of partial differential equations given by:

(129) $$\frac{\partial u}{\partial \bar{z}_j} = \alpha_j, \quad j = 1, \ldots, n.$$

where the α_j satisfy the compatability conditions:

(130) $$\frac{\partial \alpha_j}{\partial \bar{z}_p} - \frac{\partial \alpha_p}{\partial \bar{z}_j} = 0.$$

Given $\alpha_1, \ldots, \alpha_n$ on Ω satisfying (130) we wish to find u satisfying (129).

More generally, in the standard notation of differential forms we let $\mathcal{A}^{p,q}$ denote all α expressed by:

(131) $$\alpha = \sum_{\substack{1 \le i_1 < \ldots < i_p \le n \\ 1 \le j_1 < \ldots < j_q \le n}} \alpha_{i_1 \ldots i_p j_1 \ldots j_q} dz_{i_1} \wedge \ldots \wedge dz_{i_p} \wedge d\bar{z}_{j_1} \wedge \ldots \wedge d\bar{z}_{j_q}$$

and, as usual we define $\bar{\partial} : \mathcal{A}^{p,q} \to \mathcal{A}^{p,q+1}$ by:

(132) $$\bar{\partial}\alpha = \Sigma\, \bar{\partial}(\alpha_{i_1 \ldots i_p j_1 \ldots j_q}) \wedge dz_{i_1} \wedge \ldots \wedge d\bar{z}_{j_q}.$$

The problem then is given $\alpha \in \mathcal{A}^{p,q}(\Omega)$ satisfying:

(133) $$\bar{\partial}\alpha = 0 ,$$

to find $\psi \in \mathcal{A}^{p, q-1}(\Omega)$ such that

(134) $$\bar{\partial}\psi = \alpha .$$

Here we are mainly interested in the regularity properties of ψ. For simplicity most of our discussion will concern the case $p = 0$, $q = 1$.

Suppose that $\Omega \subset \mathbb{C}^2$ and that the origin lies on $b\Omega$. Further, suppose that there is a neighborhood U of the origin so that $U \cap \bar{\Omega}$ is strictly convex and suppose further that $U \cap \bar{\Omega}$ is contained in the set $\{(z_1, z_2) \mid \mathrm{Re}(z_2) \leq 0\}$. Then the function z_2^{-1} is holomorphic on $U \cap \Omega$ and is smooth on $U \cap \bar{\Omega} - \{(0,0)\}$, but it cannot be extended as holomorphic function to any neighborhood of $(0,0)$. Now suppose that on Ω we can always solve the system (129) provided α satisfies (130) and suppose also that when α is in $C^\infty(\bar{\Omega})$ then there is a solution $u \in C^\infty(\bar{\Omega})$. Let $\rho \in C_0^\infty(U)$ such that $\rho = 1$ or a neighborhood V of the origin. Set

(135) $$\alpha = \bar{\partial}\left(\frac{\rho}{z_2}\right) = \frac{\bar{\partial}\rho}{z_2} ,$$

then, since $\bar{\partial}\rho = 0$ in V, we see that $\alpha \in C^\infty(\bar{\Omega})$; so, by assumption, there exists $u \in C^\infty(\bar{\Omega})$ with $\bar{\partial}u = \alpha$ (observe also that $\bar{\partial}\alpha = 0$). Now, if the compliment of Ω has a bounded component, denoted by Θ then Hartog's theorem states that every holomorphic function on Ω can be extended to a holomorphic function on $\Omega \cup \Theta$. In particular, if $(0,0) \in \Theta$ then every holomorphic function in Θ can be extended past $(0,0)$. Let

(136) $$h = u - \frac{\rho}{z_2},$$

then h is holomorphic on Ω but clearly h cannot be extended past (0, 0) since u is smooth on $\bar{\Omega}$. Thus we see that (129) cannot be solved (with smooth solutions) on a domain with "holes".

If Ω is convex (or more generally psuedo-convex, as will be explained below) then the equations (129) always have a solution (see [11]). The question that we will discuss here is the regularity of solution. Suppose $\Omega \subset \mathbb{C}^n$ is convex and that in a neighborhood U' of the origin we have:

(136) $$U' \cap \Omega = \{(z_1, z_2) \in U' \mid \mathrm{Re}(z_2) < 0\}.$$

Choose a subneighborhood U of (0, 0) such that $\bar{U} \subset U'$ and a function ρ as above. Again we define α by (135). Now suppose that there is some solution u of (129) which is regular where α is regular. Then, in particular u is regular in $(U'-U) \cap \bar{\Omega}$ and in $V \cap \bar{\Omega}$. Let h be the holomorphic function defined by (136). For δ positive, we evaluate h on the line $z_2 = -\delta$ so that

(137) $$h(z_1, -\delta) = u(z_1, -\delta) + -\frac{\rho(z_1, -\delta)}{\delta}.$$

For given positive numbers δ_0 and R we define the set S by:

(138) $$S = \{(z_1, -\delta) \mid |z_1| = R, \quad 0 < \delta < \delta_0\}.$$

We assume that U' and U were chosen so that $S \subset U' - U$. Since u is regular in $V \cap \bar{\Omega}$ and in $(U'-U) \cap \bar{\Omega}$ we know that, in particular, n is bounded independently of δ or S and the set $\{(0,-\delta) | 0 < \delta < \delta_0\}$. Now evaluating at $(0,-\delta)$ we see that it behaves like $\frac{1}{\delta}$ there and that on the circle $(z_1, -\delta)$ with $|z_1| = R$ with $0 < \delta < \delta_0$, h is bounded independently of δ. This is impossible since $h(0,-\delta)$ is the average of $h(z_1, -\delta)$ with $|z_1| = R$. Hence, in this example, we see that no solution of (129) sing supp(u) $\subset$ sing supp(α). Here by sing supp stands for singular support and sing supp(u) is the subset of $\overline{\Omega}$ such that if $(z_1, z_2) \notin$ sing supp(then there exists a neighborhood U of (z_1, z_2) such that u is C^∞ or $U \cap \overline{\Omega}$. The same type of argument as above proves the following result.

139. Proposition. If $\Omega \subset \mathbb{C}^n$ is convex and if there is a germ of a holomorphic curve contained in $b\Omega$ then there exists $\alpha \in \mathcal{A}^{0,1}$ with $\overline{\partial}\alpha = 0$ such that whenever $\overline{\partial}u = \alpha$ we have sing supp(u) $\not\subset$ sing supp(α).

Here, as usual, a germ of a holomorphic curve means a germ of a complex analytic variety of dimension one.

In studying existence and regularity for $\overline{\partial}$ on domains with boundary a crucial role is played by the Levi form. Given a point $x_0 \in b\Omega$ we define an (n-1)-dimensional subspace of $\mathbb{C}^n$ by

$$\sum_1^n \zeta_j r_{z_j}(x_0) = 0 . \tag{140}$$

The Levi form is the quadratic form which operates on this subspace, defined by:

$$\sum r_{z_i \overline{z}_j}(x_0)\zeta_i\overline{\zeta}_j \tag{141}$$

We say that Ω is pseudo-convex at x_0 if the Levi form at x_0 is now negative and Ω is strictly pseudo-convex at x_0 if the Levi form at x_0 is positive definite. To define the Levi-form invariantly we define a sub-bundle of the complex tangent bundle to $b\Omega$, denoted by $T_b^{1,0}$. The fiber of $T_b^{1,0}$ at x_0 consists of all the vector at x_0 of the form $L = \Sigma \zeta_j \frac{\partial}{\partial z_j}$ such that (140) holds. Another way of expressing this is by $L(r) = 0$. Then the Levi form acts on $T_b^{1,0}$ by sending L into $<\partial\bar{\partial}r, L \wedge \bar{L}>$, where $\bar{L} = \Sigma \bar{\zeta}_j \frac{\partial}{\partial \bar{z}_j}$, here $<,>$ denotes contraction. It follows from a Lie algebra identity, due to Cartan, that if L and L' are two vector fields with values in $T_b^{1,0}$, then

$$(142) \qquad <\partial\bar{\partial}r, L \;\; \bar{L}'> = <\partial r, [L, \bar{L}']>$$

where $[L, \bar{L}'] = L\bar{L}' - \bar{L}'L$.

Given $x_0 \in b\Omega$ in a neighborhood U of x_0 we wish to define a special basis for the 1-form, in terms of which the quadratic form Q, that corresponds to $\bar{\partial}$, is easier to analyze. Let $\omega^n = g\partial r$, where g is the function chosen so that $|\omega^n| = 1$ at each point of U, i.e.

$$(143) \qquad g = (2\Sigma |r_{z_j}|^2)^{-\frac{1}{2}} .$$

Let $\omega^1, \ldots, \omega^n$ be an orthonormal basis of the (1, 0)-forms at each point of U. Then $\bar{\omega}^1, \ldots, \bar{\omega}^n$ is an orthonormal basis of (0,1)-forms. Let $L_1, \ldots, L_n$ be the dual basis of $\omega^1, \ldots, \omega^n$. Observe that:

$$(144) \qquad L_j(r) = <L_j, \partial r> = \frac{1}{g} <L_j, \omega^n> = \frac{1}{g} \delta_{jn} .$$

Thus, on $b\Omega$ the vector fields $L_1, \ldots, L_{n-1}$ have values in $T_b^{1,0}$ and similarly $\bar{L}_1, \ldots, \bar{L}_{n-1}$ have values in $T_b^{0,1}$. We define the vector field T on U by:

(145) $$T = L_n - \bar{L}_n .$$

It follows from (144) that $T(r) = 0$ and hence the restrictions of $L_1, \ldots, L_{n-1}$, $\bar{L}_1, \ldots, \bar{L}_{n-1}, T$ from a local basis of the tangent space of $b\Omega$. If, $i, j < n$ then the vector field $[L_i, \bar{L}_j]$ is tangent to $b\Omega$ and so on $b\Omega$ it can be expressed by:

(146) $$[L_i, \bar{L}_j] = c_{ij} T + \sum_1^{u-1} a_{ij}^k L_k + \sum_1^{u-1} b_{ij}^k \bar{L}_k .$$

From (142) it then follows that the Levi form in terms of this basis is given by $g^{-1} c_{ij}$. We will renormalize r so that $g = 1$ on $b\Omega$ then the Levi form is e_{ij} with respect to the basis $L_1, \ldots, L_{n-1}$.

Relative to this basis we represent a $(0,1)$-form by $\varphi = \Sigma \varphi_j \bar{\omega}^j$, then we have:

(147) $$\bar{\partial} u = \Sigma (\bar{L}_j u) \bar{\omega}^j$$

and

(148) $$\bar{\partial}\varphi = \Sigma \bar{\partial}\varphi_j \wedge \bar{\omega}^j = \sum_{k<j} (\bar{L}_k \varphi_j - \bar{L}_j \varphi_k) \bar{\omega}^j \wedge \bar{\omega}^k + \ldots ,$$

when the dots represent terms which are linear combinations of the φ_j's (they do not involve any derivatives of the φ_j).

Lecture 6.

The following theorem is proven in [12] and [13].

149. Theorem. If Ω is pseudo-convex and if α is a $(0,1)$-form with components in $C^\infty(\bar{\Omega})$ and if $\bar{\partial}\alpha = 0$ then there exists a $u \in C^\infty(\bar{\Omega})$ such that $\bar{\partial}u = \alpha$.

This is a theorem of global regularity it is proven by using a of Q with appropriate "weight functions". Here we will concentrate on local regularity by means of establishing subelliptic estimates. From proposition 139 we know that local regularity sequences additional conditions. We will now consider the form Q on $(0,1)$-forms associated with $\bar{\partial}$. We have:

$$Q(\varphi,\psi) = (\bar{\partial}\varphi, \bar{\partial}\psi) + (\bar{\partial}^*\varphi, \bar{\partial}^*\psi) + (\varphi,\psi) \tag{150}$$

for $\varphi,\psi \in \mathcal{D}$ and $\mathcal{D} = (\mathrm{Dom}\,\bar{\partial}) \cap (\mathrm{Dom}\,\bar{\partial}^*)$. Integration by parts shows that

$$(\bar{L}_j u, v) = -(u, L_j v) + (u, g_j v) + \delta_{jn} \int_{b\Omega} u\bar{v}\,dS \tag{151}$$

when $u, v \in C_0^\infty(U \cap \bar{\Omega})$. Hence, on $U \cap \bar{\Omega}$, we have

$$\bar{\partial}^*\varphi = -\sum L_j\varphi_j + \sum g_j\varphi_j \tag{152}$$

and

$$\dot{\mathcal{D}} = \{\varphi = \sum \varphi_j \bar{\omega}^j,\ \varphi_j \in C^\infty(\bar{\Omega}) \mid \varphi_n = 0 \text{ on } b\Omega\}\,. \tag{153}$$

154. Proposition. If $x_0 \in b\Omega$ and if the Levi-form is non-negative at x_0 then there exists a neighborhood U of x_0 and constant $c > 0$ such that

$$\|\varphi_{\bar{z}}\|^2 + \sum_1^{n=1} \int_{b\Omega} c_{ij}\varphi_i\bar{\varphi}_j dS \leq CQ(\varphi,\varphi), \tag{155}$$

for all $\varphi \in \dot{\mathcal{D}}_U$. Here $\dot{\mathcal{D}}_U = \{\varphi \in \dot{\mathcal{D}} \mid \operatorname{supp}(\varphi) \subset U \cap \bar{\Omega}\}$ and $\|\varphi_{\bar{t}}\|^2 = \sum_1^n \|L_j\varphi_k\|^2$.

The following is a consequence of theorem 103.

156. Corollary. If the Levi-form is strictly positive definite at x_0 then Q is subelliptic at x_0 with $\varepsilon = \frac{1}{2}$.

The inequality (155) is derived from the following identity:

$$Q(\varphi,\varphi) = \|\varphi_{\bar{z}}\|^2 + \sum_1^{n-1} \int_{b\Omega} c_{ij}\varphi_i\bar{\varphi}_j\, ds + O(\|\varphi_{\bar{z}}\|\,\|\varphi\| + \|\varphi\|^2) \tag{157}$$

which is valid for all $\varphi \in \dot{\mathcal{D}}_U$ without any assumptions on the Levi form. We will give a brief outline of the proof of (157). From (148) we have:

$$\|\bar{\partial}\varphi\|^2 = \sum_{k<j} \|\bar{L}_k\varphi_j - \bar{L}_j\varphi_k\|^2 + O(\|\varphi_{\bar{z}}\|\,\|\varphi\| + \|\varphi\|^2) \tag{158}$$

$$= \|\varphi_{\bar{z}}\|^2 - \sum_{k,j} \operatorname{Re}(\bar{L}_k\varphi_j, \bar{L}_j\varphi_k) + O(\ldots)$$

Now by integrating twice by parts we have:

$$(\bar{L}_k\varphi_j, \bar{L}_j\varphi_k) = (L_j\varphi_j, L_k\varphi_k) - \sum_1^{n-1} (c_{jk}T\varphi_j, \varphi_k) \tag{159}$$

$$+ O(\|\varphi_{\bar{z}}\|\,\|\varphi\| + \|\varphi\|^2).$$

Since L_j for $j < n$ and tangential and since $\varphi_n = 0$ on $b\Omega$, no boundary terms appear in the above. Combining (158) and (159) with the expression for $\bar{\partial}^*$ in (152) we obtain the desired identity (157).

To investigate the subellipticity of Q we follow the method described in [14].

160. Definition. If $x_0 \in b\Omega$ we define the set of subelliptic multipliers at x_0 denoted by I to be the set of germs of functions at x_0 which satisfy the following condition. $f \in I$ if and only if there exists a neighborhood U of x_0 and constants $\varepsilon > 0$, $c > 0$ such that

(161) $$\|f\varphi\|_{\varepsilon}^2 \leq CQ(\varphi, \varphi) \text{ for all } \varphi \in \dot{\mathcal{D}}_U .$$

162. Definition. If $x_0 \in b\Omega$ we define the set of subelliptic functionals at x_0 denoted by F to be the set of n-tuples of germs of functions at x_0 which satisfy the following. $v = (v_1, \ldots, v_n) \in F$ if and only if there exists a neighborhood U of x_0 and constants $\varepsilon > 0$, $c > 0$ such that

(163) $$\|\sum_1^n v_i\varphi_i\|^2 \leq CQ(\varphi, \varphi) \text{ for all } \varphi \in \dot{\mathcal{D}}_U .$$

164. Theorem. If Ω is pseudo-convex at $x_0 \in b\Omega$ then I and F have the following properties:

(a) I is an ideal.

(b) $I = \sqrt[\mathbb{R}]{I}$, when $\sqrt[\mathbb{R}]{I}$ denotes the real radical of I and is defined by $h \in \sqrt[\mathbb{R}]{I}$ if and only if there exists and integer and $f \in I$ such that $|h|^m \leq |f|$.

(c) If $f \in I$ then $(L_1 f, L_2 f, \ldots, L_n f) \in F$.

(d) If $v^{(1)}, \ldots, v^{(n)} \in F$ then $\det v_j^{(i)} \in I$.

(e) $r \in I$.

(f) $(c_{i_1}, c_{i_2}, \ldots, c_{i_{n-1}}, 0) \in F$ for $i = 1, \ldots, n-1$.

To prove that Q is subelliptic at x_0 is equivalent to proving that $1 \in I$. Property (a) follows from

$$Q(f\varphi, f\varphi) \leq \max(f)^2 Q(\varphi, \varphi) + \text{const.} \ \|\varphi\|^2 \tag{165}$$

which is immediate. To prove (b) let $h \in \sqrt[\mathbb{R}]{I}$, then it suffices to show that for some δ we have:

$$\|\Lambda_t^\delta (h\varphi)\|^2 \leq CQ(\varphi, \varphi) \tag{166}$$

since then the fact that $b\Omega$ is non-characteristic will imply that $\|h\varphi\|_\delta^2 \leq C'Q(\varphi, \varphi)$. We have

$$\begin{aligned} \|\Lambda_t^\delta (h\varphi)\|^2 &= (|h|^2\varphi, \rho^{2\delta}\varphi) \leq \|\Lambda_t^{2\delta}(h^2\varphi)\|^2 + \text{const.} \ \|\varphi\|^2 \\ &\leq \|\Lambda_t^{2^k\delta}(h^{2k}\varphi)\|^2 + \text{const.} \ \|\varphi\|^2 , \end{aligned} \tag{167}$$

when $\delta^{2\delta}$ is a tangential pseudo-differential operator of order 2δ. Then, if $2^k\delta \leq \varepsilon$ we have

$$\|\Lambda_t^{2^k\delta}(h^{2k}\varphi)\|^2 \leq \|\Lambda_t^\varepsilon (h^{2k}\varphi)\|^2 \leq \int |h|^{4k} |\Lambda_t^\varepsilon \varphi|^2 dV + e\|\varphi\|^2 \tag{168}$$

and if $|h|^{2k} \leq |f|$ then

(169) $$\int |h|^{4k} |\Lambda_t^{\varepsilon}\varphi\|^2 dV \leq \|\Lambda_t^{\varepsilon}(f\varphi)\|^2 + \text{const. } \|\varphi\|^2 ,$$

which concludes the proof of (b).

Property (c) follows from the observation that $\|\bar{\partial}^*\varphi\| \leq Q(\varphi, \varphi)$ and thus:

(166) $$\|\sum_1^n L_i\varphi_i\|^2 \leq CQ(\varphi, \varphi) .$$

By non-characteristicity we have

(167) $$\|\Sigma (L_i f)\varphi_i\|_\delta^2 \leq \text{const. } (\|\Lambda_t^\delta \Sigma (L_i f)\varphi_i\|^2 + Q(\varphi, \varphi)) .$$

Then

(168) $$\|\Lambda_t^\delta \Sigma (L_i f)\varphi_i\|^2 = (\Sigma (L_i f)\varphi_i, S^{2\delta}\varphi) = -(f\Sigma L_i\varphi_i, S^{2\delta}\varphi) - \Sigma(f\varphi_i, S^{2\delta}\bar{L}_i\varphi) + \text{const. } \|\varphi\|^2$$

$$\leq \text{const. } \|f\varphi\|_{2\delta}(\|\Sigma L_i\varphi_i\| + \|\varphi_{\bar{z}}\| + \|\varphi\|) + \text{const. } \|\varphi\|^2$$

So choosing $\delta \leq \frac{1}{2}\varepsilon$ proves (c).

Property (d) is a consequence of the inequality

(169) $$(\det A_{ij})\Sigma|\zeta_k|^2 \leq \text{const. } \Sigma A_{ij}\zeta_i\bar{\zeta}_j ,$$

whenever A_{ij} is non-negative, in this case we have $A_{ij} = \sum_k v_i^{(k)}\bar{v}_j^{(k)}$.

Property (e) follows immediately by replacing φ by $r\varphi$ in (157); in fact, we obtain $\|r\varphi\|_1^2 \leq CQ(\varphi,\varphi)$.

To prove property (f) we first establish the following inequality:

$$(170) \qquad \left|\left(\sum_{j=1}^{n-1} c_{ij}\varphi_j, D_t S^0 \psi_i\right)\right| \leq \text{const.}\ (Q(\varphi,\varphi)+Q(\psi,\psi))$$

for all $\varphi, \psi \in \dot{\mathcal{D}}_U$; where D_t is any tangential vector field on U (i.e. a vectorfield such that $D_t r = 0$ on $b\Omega$) and S^0 is a tangential pseudo-differential operator of order 0 on U. The inequality (169) is established by expressing D_t as a combination of $L_1, \ldots, L_{n-1}, \bar{L}_1, \ldots, \bar{L}_{n-1}$ and T and reasoning as in the proof of proposition 154. To prove property (f) we set $\psi_i = \sum_k c_{ik}\varphi_k$ and we prove that $Q(\psi,\psi) \leq \text{const.}\ Q(\varphi,\varphi)$. We then let $D_k = \frac{\partial}{\partial t_k}$ where $t_1, \ldots, t_{2n-1}$, r is a boundary coordinate system and apply (170) to $D_t = D_k$ and $S^\delta = D_k \Lambda_t^{-1}$ then summing over k we have

$$(171) \qquad \left|\left(\psi_i, \sum_1^{2k-1} D_k^2 \Lambda_t^{-1} \psi_i\right)\right| \leq CQ(\varphi,\varphi)$$

and since

$$(172) \qquad \|\Lambda_t^{\frac{1}{2}} u\|^2 = (u, \Lambda_t^1 u) = \left(u, \sum_1^{2k-1} D_k^2 \Lambda_t^{-1} u\right) + (u, \Lambda_t^{-1} u)$$

$$\leq \left|\left(u, \sum_1^{2n-1} D_k^2 \Lambda_t^{-1} u\right)\right| + \text{const.}\ \|u\|^2 .$$

This establishes property (f).

Lecture 7

We will now use the properties of I given in the above theorem to construct an increasing sequence of ideals contained in I. If G is a set of n-vectors we will denote by det G the set consisting of all determinants of $n \times n$ matrices whose rows are vectors in G. Let F_0 consist of the vectors $(c_{i_1}, \ldots, c_{in-1}, 0)$ for $i=1,\ldots,n-1$ and $(0,\ldots,0,1)$. Set $I_1 = \sqrt[\mathbb{R}]{(r, \det F_0)}$, where $(r, \det F_0)$ denotes the ideal generated by the set $\{r, \det F_0\}$. Inductively, we let $F_k = F_{k-1} \cup \{(L_1 g, \ldots, L_n g) \mid g \in I_k\}$ and $I_{k+1} = \sqrt[\mathbb{R}]{(I_k, \det F_k)}$. We then have $I_k \subset I_{k+1} \subset I$. Now note that $F_k = F_0 \cup \{(L_1 g, \ldots, L_n g) \mid g \in I_k\}$. We denote by $\mathcal{V}(I_k)$ the variety of I_k, that is all x near x_0 such that $f(x) = 0$ for all $f \in I_k$. To get a geometrical picture of the process of the process of passing from I_k to I_{k+1} we define for $x \in V(I_k)$ the Zariski tangent space of I_k at x_1 denoted by $Z_x^{1,0}(I_k)$, as follows:

$$Z_x^{1,0}(I_k) = \{L = \Sigma \zeta_j L_j \mid L(f)]_x = 0\} . \tag{173}$$

Further we define the null space of the Levi form $\mathcal{N}_x$ by:

$$\mathcal{N}_x = \{L = \Sigma \zeta_j L_j \mid \Sigma c_{ij}(x)\zeta_i\bar{\zeta}_j = 0\}. \tag{174}$$

175. <u>Proposition.</u> If x is sufficiently near x_0 then $x \in \mathcal{V}(I_{k+1})$ if and only if $x \in \mathcal{V}(I_k)$ and $Z_x^{1,0}(I_k) \cap \mathcal{N}_x \neq \{0\}$.

This proposition is an immediate consequence of the definitions. Since $x \in \mathcal{V}(I_{k+1})$ is equivalent to $x \in \mathcal{V}(I_k)$ and all determinants in F_k vanish at x; this means that there exists $(\zeta_1, \ldots, \zeta_n) \neq 0$ which satisfies

$$\Sigma a_j \zeta_j = 0 ,$$

whenever $(a_1, \dots, a_n) \in F_k$ and this means that

$$\sum_i c_{ij} \zeta_i = 0 \qquad j=1, \dots, n-1$$

and

$$\Sigma (L_i g) \zeta_i = 0 \quad \text{for} \quad g \in I_k$$

which is equivalent to $\Sigma \zeta_i L_i \in Z_x^{1,0}(I_k) \cap \mathcal{N}_x$.

If we now assume that r is real analytic in a neighborhood of x_0 then we can apply the theory of ideals of real analytic functions to the situation. We restrict the definitions of all the ideals introduced above to ideals of germs of real analytic functions at x_0. Then we can apply a very important result of Lojasiewicz (see [15]) which gives the "Nullstellen Satz" for real analytic functions. Namely, let J be an ideal of germ of real analytic functions at x_0 and let $\mathcal{I}_{x_0}(\mathcal{V}(J))$ be the ideal of germs of analytic functions at x_0 that vanish on $\mathcal{V}(J)$, then we have

(176) $$\sqrt[\mathbb{R}]{J} = \mathcal{I}_{x_0}(\mathcal{V}(J)) .$$

Another important theorem that is useful here is the coherence theorem of Cartan. This theorem says that if V is a germ of real-analytic variety at x_0 and if $\mathcal{I}_x(V)$ denotes the ideal of germs of real analytic functions at x which vanish on V, then there exists a neighborhood U of x_0 and a set $S \subset V \cap U$ such that S is open and dense in $V \cap U$ and such that for each $x \in S$ there exists finitely many elements of $\mathcal{I}_{x_0}(V)$ that generate $\mathcal{I}_x(V)$.

177. Definition. If V is a germ of a real analytic variety at $x_0 \in b\Omega$ and $V \subset b\Omega$ we define the holomorphic dimension of V by

(178)
$$\text{hol dim}(V) = \min_{x \in V} \dim(Z_x^{1,0} \mathcal{J}_{x_0}(V)) \cap \mathcal{N}_x) .$$

179. Theorem. If $x_0 \in b\Omega$, if Ω is pseudo-convex, if r is analytic in a neighborhood of x_0 and if there exists a neighborhood of U of x_0 such that $U \cap b\Omega$ does not contain any germs of real analytic varieties of positive holomorphic dimension than Q is subelliptic at x_0.

Let S be the set of all $x \in b\Omega$ such that there is a germ of a real analytic variety V with $x \in V \subset b\Omega$ and $\text{hol dim}(V) > 0$. We will outline a proof of the fact that if r is real analytic in a neighborhood of x_0 and if Ω is pseudo-convex then there exists a neighborhood U of x_0 such that

(180)
$$U \cap \bigcap_k \mathcal{V}(I_k) \subset \bar{S}$$

The theorem then follows from (180) since it shows that when $S = \phi$ then $\mathcal{V}(I) = \phi$ and hence $1 \in I$. To prove (180) we will show that if $U \cap \mathcal{V}(I_k) - \bar{S}$ is not empty then it has an open dense subset which is not contained in $V(I_{k+1})$. Suppose x is a simple point of $\mathcal{V}(I_k)$ and suppose that $\mathcal{J}_x(V(I_k))$ is generated by a finite number of elements in I_k then x has a neighborhood $W \subset \mathcal{V}(I_k)$ so that each point in W has the same properties and so that $W \cap \bar{S} = \phi$. Now suppose that W has an open subset W' such that $W' \subset \mathcal{V}(I_{k+1})$ then for $y \in W'$ we have

$$Z_y^{1,0}(I_k) \cap \mathcal{N}_y \neq \{0\} \text{ and } Z_y^{1,0}(I_k) = Z_y^{1,0}(\mathcal{J}_{y_0}(W'))$$

for any $y_0 \in W'$ hence hol dim $W' > 0$ and so $W' \subset S$ which is a contradiction.

The following theorem has recently been proved by Diedrich and Fornaess (see [16]).

181. Theorem. If Ω is pseudo-convex, if $x_0 \in b\Omega$, if r is real analytic in a neighborhood U of x_0 and if $V \subset U \cup b\Omega$ is a germ of a real analytic variety then there exists a germ of complex analytic variety $W \subset U \cup b\Omega$ such that hol dim V = dim W.

This theorem is very deep, in fact W is not in general a subset of V and the structure defined by $Z_x^{1,0}(\mathcal{J}(V)) \cap \mathcal{N}_x$ is not integrable.

Combining Theorems 181 and 179 we obtain the following.

182. Theorem. If Ω is pseudo-convex, $x_0 \in b\Omega$, r is real analytic in a neighborhood of x_0 and if there exist a neighborhood U of x_0 such that $U \cap b\Omega$ does not contain any germs of holomorphic curves then Q is subelliptic at x_0.

In view of Proposition 139 we have

183. Corollary. If Ω is convex, $x_0 \in b\Omega$, and r is real analytic in a neighborhood of x_0 then Q is subelliptic at x_0 if and only if there exists a neighborhood U of x_0 such that $U \cap b\Omega$ does not contain germs of holomorphic curves.

Now consider the operator $\bar{\partial} : \mathcal{A}^{p,q-1} \to \mathcal{A}^{p,q}$. This leads to a quadratic form Q defined in a space $\mathcal{D}^{p,q} \subset \mathcal{A}^{p,q}$ again by formula (150). In terms of the basis $\omega^1, \ldots, \omega^n$ if we express $\varphi \in \mathcal{A}^{p,q}$ by

$$\varphi = \sum_{\substack{i_1 < \cdots < i_p \\ j_1 < \cdots < j_q}} \varphi_{i_1 \ldots i_p j_1 \ldots j_p} \omega^{i_1} \wedge \cdots \wedge \omega^{i_p} \wedge \bar{\omega}^{j_1} \wedge \cdots \wedge \bar{\omega}^{j_q},$$

then $\mathcal{D}^{p,q}$ is characterized by $\varphi \in \mathcal{D}^{p,q}$ if and only if $\varphi_{i_1 \ldots i_p j_1 \ldots j_q} = 0$ on $b\Omega$ when $j_q = n$. The procedure for studying subellipticity of Q can be generalized to this case. Given $x_0 \in b\Omega$ we define the ideal I^q for Q on (p, q)-forms the same way as I was defined for $(0, 1)$-forms (it is clear that I^q is independent of p). Similarly we define the set F^q of subelliptic functionals to be vectors $(g_1, \ldots, g_n)$ such that

$$\sum_{A, C} \Big\| \sum_{B-j=C} \sigma(j, B) g_j \varphi_{AB} \Big\|_\epsilon^2 \leq \text{const. } Q(\varphi, \varphi) \tag{184}$$

for all $\mathcal{D}_U^{p,q}$ as in previous definitions, where A, B and C range over ordered p-tuples, q-tuples and $(q-1)$-tuples, respectively and

$$\sigma(j, B) = \begin{cases} 0 & \text{if } j \notin B \\ (-1)^k & \text{if } j = b_k \end{cases} . \tag{185}$$

Then there are properties of Theorem 164 with exception of (d) hold as rotated (i.e. replacing I with I^q and F with F^q). The property (d) is generalized by

$(d)_q$ If $v^{(1)}, \ldots, v^{(n-q+1)} \in F^q$ then $\det_{(n-q+1)}(v_j^{(i)}) \subset I^q$, when $\det_{(n-q+1)}(v_j^{(i)})$ denotes the set of determinants of all $(n-q+1) \subset (n-q+1)$ minors of the metrix $(v_j^{(i)})$.

Theorem 182 then generalizes to the following statement.

186. Theorem. If Ω is pseudo-convex, $x_0 \in b\Omega$, or real analytic in a neighborhood of x_0 and if there exists a neighborhood U of x_0 such that $U \cap b\Omega$ does not contain any germs of complex analytic varieties of dimension greater or equal to q then Q on $\dot{\mathcal{D}}^{p,q}$ is subelliptic at x_0.

For the case $q = n-1$ the following criterior is easy to see.

187. Proposition. Suppose Ω is pseudo-convex, $x_0 \in b\Omega$ and r analytic in a neighborhood of x_0. Let $\lambda = \sum_1^{n-1} c_{ii}$ be the trace of the Levi-form. Then there exists a neighborhood U of x_0 such that no germs of $(n-1)$-dimensional complex-analytic varieties lie in $U \cap b\Omega$ if and only if there exists germs of vector fields $A_1, \ldots, A_k$ at x_0 with values in $T_b^{1,0} + T_b^{0,1}$ such that $A_1, \ldots, A_k(\lambda) \nmid 0$.

In fact in this case one can also obtain necessary and sufficient conditions for subellipticity with $r \in C^\infty$.

188. Theorem. If Ω is pseudo-convex and if $x_0 \in b\Omega$ then Q on $\mathcal{D}^{p,n-1}$ is subelliptic at x_0 if and only if there exist germs of vector fields $A_1, \ldots, A_k$ at x_0 with values in $T_b^{1,0} + T_b^{0,1}$ such that $A_1, \ldots, A_k(\lambda) \nmid 0$ where λ is trace of the Levi form.

The proof of this theorem is along the lines of [17] for sufficiency and [18] for necessity.

Lecture 8

One of the important consequences of subelliptic estimates is that the Bergman projection operator is pseudo-local. Let $\mathcal{H}$ denote the space of holomorphic functions on Ω which are square integrable and $B : L_2(\Omega) \to \mathcal{H}$ denote orthogonal projection. Now consider the operator on (0, 1)-forms

(189) $$\Box = \bar{\partial}\bar{\partial}^* + \bar{\partial}^*\bar{\partial}$$

with $\mathrm{Dom}(\Box) = \{\varphi \in \mathcal{D} \mid \bar{\partial}\varphi \in \mathrm{Dom}(\bar{\partial}^*)$ and $\bar{\partial}^*\varphi \in \mathrm{Dom}(\bar{\partial})\}$. This operator for general systems was discussed in Lecture 5.

We denote the null space of $\Box$ by $\mathcal{H}^{0,1}$ and as in Lecture 5 we observe that

(190) $$\mathcal{H}^{0,1} = \{\varphi \in \mathcal{D} \mid \bar{\partial}\varphi = 0, \ \ \bar{\partial}^*\varphi = 0\}.$$

If we denote by $H : L_2^{0,1}(\Omega) \to \mathcal{H}^{0,1}$ the orthogonal projection of square-integrable (0, 1)-forms onto $\mathcal{H}^{0,1}$ then we have the following result.

191. <u>Proposition.</u> If the range of $\Box$ is closed then there exists a unique bounded operator $N : L_2^{0,1}(\Omega) \to L_2^{0,1}(\Omega)$ such that

(192) $$\alpha = \Box N\alpha + H\alpha .$$

and

(193) $$HN = 0 .$$

Furthermore, if $f \in L_2(\Omega)$ and $f \in \mathrm{Dom}(\bar{\partial})$ then

(194) $$Bf = f - \bar{\partial}^* N\bar{\partial}f .$$

The operator N was constructed in Lecture 5. Applying (192) to $\alpha = \bar{\partial} f$ we have from the dimension in Lecture 6 that

(195) $$\bar{\partial} f = \bar{\partial}\,\bar{\partial}^{*} N \bar{\partial} f$$

and so $\bar{\partial} Bf = 0$, thus $Bf \in \mathcal{H}$. Further if $f \in \mathcal{H}$ then $Bf = f$ and $f \perp \mathcal{H}$ then for $h \in \mathcal{H}$ we have

(196) $$(Bf, h) = -(\bar{\partial}^{*} N \bar{\partial} f, h) = 0$$

so that $Bf = 0$. This shows that (194) defines the Bergman projections operator.

From (194) it follows that if N is pseudo-local then B is pseudo-local.

Now we take up operators induced on the boundary. Let $\mathcal{A}^{p,q}$ denote the forms of type (p, q) whose components are in $C^{\infty}(\bar{\Omega})$. We define $\mathcal{C}^{p,q} \subset \mathcal{A}^{p,q}$ by:

(197) $$\mathcal{C}^{p,q} = \{\varphi \in \mathcal{A}^{p,q} \mid \varphi \;\; \bar{\partial} r = 0 \text{ on } b\Omega\} .$$

It then follows that $\bar{\partial}(\mathcal{C}^{p,q}) \subset \mathcal{C}^{p,q+1}$ so that we have:

(198) $$\begin{array}{ccccccccc} 0 & \longrightarrow & \mathcal{C}^{p,q+1} & \longrightarrow & \mathcal{A}^{p,q+1} & \longrightarrow & \mathcal{B}^{p,q+1} & \longrightarrow & 0 \\ & & \bar{\partial}\uparrow & & \bar{\partial}\uparrow & & \bar{\partial}_b\uparrow & & \\ 0 & \longrightarrow & \mathcal{C}^{p,q} & \longrightarrow & \mathcal{A}^{p,q} & \longrightarrow & \mathcal{B}^{p,q} & \longrightarrow & 0 , \end{array}$$

where $\mathcal{B}^{p,q}$ is the quotient $\mathcal{A}^{p,q}/\mathcal{C}^{p,q}$ and it is easy to check that $\bar{\partial}_b$ is well defined. In case $p = q = 0$, $\mathcal{B}^{0,0} = C^{\infty}(b\Omega)$ and in a neighborhood

U with a special basis $\omega^1, \ldots, \omega^n$ as before we can express $\varphi \in \mathcal{B}^{0,1}$ as

(199) $$\varphi = \sum_1^{n-1} \varphi_j \bar{\omega}^j, \quad \text{with } \varphi_j \in C^\infty(U \cap b\Omega)$$

and

(200) $$\bar{\partial}_b u = \sum_1^{n-1} \bar{L}_j(u)\bar{\omega}^j \quad \text{with } u \in C^\infty(U \cap b\Omega) .$$

At each point in $b\Omega$ we have an inner product on the (p, q)-forms evaluated at that point, this induces an inner product on $\mathcal{B}^{p,q}$ so that if $\varphi, \psi \in \mathcal{B}^{p,q}$ we have $\varphi \cdot \bar{\psi} \in C^\infty(b\Omega)$ and we define

(201) $$(\varphi, \psi) = \int_{b\Omega} \varphi \cdot \bar{\psi}\, dS ,$$

where dS is the volume element in $b\Omega$. On $\mathcal{B}^{p,q}$ consider the quadratic form

(202) $$Q_b(\varphi, \psi) = (\bar{\partial}_b \varphi, \bar{\partial}_b \psi) + (\bar{\partial}_b^* \varphi, \bar{\partial}_b^* \psi) + (\varphi, \psi)$$

The following theorem then holds (see [19] and [20]).

202. Theorem. The estimate

(203) $$\|\varphi\|_{\frac{1}{2}}^2 \leq C Q_b(\varphi, \varphi) \quad \text{for all } \varphi \in \mathcal{B}^{p,q}$$

holds if and only if the Levi form has at each point of $b\Omega$ at least $\max(q+1, n-q)$ non-zero eigen-values of the same sign.

Let $\mathcal{H}_b \subset L_2(b\Omega)$ be the null space of $\bar{\partial}_b$ (i.e. the L_2-closure of

$\bar{\partial}_b$). To each element of $\mathcal{H}_b$ this corresponds a unique holomorphic function on Ω whose "boundary value" it is. We define the Szego operator S. $L_2(b\Omega) \to \mathcal{H}_b$ to be the orthogonal projection. Now if (202) holds then the operator

(204) $$\Box_b = \bar{\partial}_b \bar{\partial}_b^* + \bar{\partial}_b^* \bar{\partial}_b$$

has a closed range and hence there exists a pseudo-local operator $N_b : L_2^{0,1}(b\Omega) \to L_2^{0,1}(b\Omega)$ defined analogously with (192), (193). Then we have for $f \in L_2(b\Omega)$ and $f \in \mathrm{Dom}(\bar{\partial}_b)$

(205) $$Sf = f - \bar{\partial}_b^* N_b \bar{\partial}_b f .$$

Let $P : C^\infty(b\Omega) \to C^\infty(b\Omega)$ be a differential operator with the property that $\mathcal{H}_b$ is contained in the null space of P^*. Then, clearly, the range of P is orthogonal to $\mathcal{H}_b$ and hence

(206) $$SP = 0.$$

We wish to study the local solvability of the equation

(207) $$Pu = f ,$$

that is if f is a function defined in a neighborhood U of $x_0 \in b\Omega$ (we take $U \subset b\Omega$) we wish to find u defined on a neighborhood U' of x_0 so that (206) is satisfied on U'.

208. Proposition. If $f \in L_2^{loc}(U)$ and if $\exists$ a weak solution $u \in L_2^{loc}(U')$ of (206) then Sf is "smooth" on U", for any "neighborhood U" such that

$\bar{U}'' \subset U'$.

Here by "smooth" on U'' we mean that it belongs to the class of functions on U'' which are restrictions of Sg to U'' with $g \in L_2(b\Omega)$ and $g = 0$ on a neighborhood of $\bar{U}''$. Thus, since S is pseudo local, "smooth" means at least in $C^\infty(U')$..

To prove 208 take $\zeta \in C_0^\infty(U')$ with $\zeta = 1$ on a neighborhood of $\bar{U}''$. Let $\tilde{f} \in L_2(b\Omega)$ such that $\tilde{f} = f$ on a neighborhood of $\bar{U}''$. Let $v = \zeta u$ on U' and $v = 0$ outside of U'. Consider $g = \tilde{f} - Pv$, clearly $g = 0$ on a neighborhood of $\bar{U}''$ hence $Sg = S\tilde{f}$ is "smooth". Observe that the smoothness of $S\tilde{f}$ on U'' is independent of the values of $\tilde{f}$ outside of U''.

Recently F. Treves proved the following result (see [21]).

209. Theorem (Treves). If Ω satisfies the hypothesis of Theorem 201, if the Levi-form is non-degenerate at every point of $b\Omega$ and if further the function r is real analytic then the operator $\Box_b$ is analytic hypoelliptic.

This then implies that the operator N_b is pseudo-local in the analytic sense, i.e. $N_b\varphi$ is analytic on any open set on which φ is analytic. In particular, if the conditions are satisfied for $q = 1$ (i.e. at least $\max(2, n-1)$ eigenvalues of the same sign and non-degeneracy of Levi form) thus the operator S is analytic pseudo local.

Now take $f \in \mathcal{H}_b$ to be the boundary value of a holomorphic function which is not real analytic in a neighborhood of x_0 then the equation (207) does not have a solution in any neighborhood of x_0.

The following are two examples of P that satisfy (206). First let P be the differential operator whose adjoints can be locally expressed by:

(210) $$P^* = \sum_{1}^{n-1} \zeta_j \bar{L}_j .$$

As a second example we take P to be

(211) $$P = \bar{\partial}_b^* \bar{\partial}_b .$$

In this example we have the following result.

212. Proposition. If N_b on (0,1) forms is analytic pseudo-local, if P is given by (211) and if f is a function defined in a neighborhood of $x_0 \in b\Omega$ then the condition Sf is analytic in some neighborhood of x_0 is necessary and sufficient for the solvability of (207).

Proof: The necessity has been proven above. The sufficiency follows by first solving the gloval problems. That is, take $\tilde{f} \in L_2(b\Omega)$ so that $\tilde{f} = f$ in a neighborhood of x_0 we wish to find v satisfying

(213) $$f = Pv + Sf$$

it is easy to check that if v is given by

(214) $$v = \bar{\partial}_b^* N_b^2 \bar{\partial}_b f$$

then v satisfies (213).

Since Sf is analytic then, by use of the theorem of Cauchy-Kowalevski there exists an analytic function w defined in a neighborhood of x_0 such that Pw = Sf. Then the solution of (207) is u = v + w.

The first example of the phenomenon of non-solvability is the famous Lewy equation (see [22]). This equation is given by the operator

$P : C^\infty(\mathbb{R}^3) \to C^\infty(\mathbb{R}^3)$ defined by:

$$P = \frac{\partial}{\partial z} + i\bar{z}\frac{\partial}{\partial t}, \tag{215}$$

where $z = x + iy$ and x, y, t then coordinates on $\mathbb{R}^3$. This operator corresponds to the boundary of the domain $\Omega \subset \mathbb{C}^2$ given by:

$$\Omega = \{(z_1, z_2) \in \mathbb{C}^2 \,|\, \mathrm{Im}\, z_2 > |z_1|^2\} \tag{216}$$

then

$$b\Omega = \{(z_1, z_2) \in \mathbb{C}^2 \,|\, \mathrm{Im}\, z_2 = |z_1|^2\}.$$

Up to multiples on $b\Omega$ there is only one operator of the form $a\frac{\partial}{\partial z_1} + b\frac{\partial}{\partial z_2}$ which is tangent to $b\Omega$, this operator is given by:

$$L = \frac{\partial}{\partial z_1} + 2i\bar{z}_1\frac{\partial}{\partial z_2}. \tag{217}$$

Consider the correspondence between $\mathbb{R}^3$ and $b\Omega$ given by $(z_1, z_2) \in b\Omega$

$$z_1 = z \text{ and } z_2 = t + i|z|^2 \tag{218}$$

Under this correspondence $P = L$. If we let volume element on $b\Omega$ be given by $dx\,dy\,dt$ then $L^* = -\bar{L}$ and hence condition (206) is satisfied in this case we can easily check that that the operator given in (211) is $-L\bar{L}$. Hence if we can show that N_b for this domain is analytic pseudo-local then we would obtain that the necessary and sufficient condition for the local solvability of $Lu = f$ is the analyticity of Sf. In this case however, the general theorems do not apply since the Levi form is 1×1 and thus cannot have two non-vanishing eigenvalues.

In this case, however, the boundary $b\Omega$ is a group and the operators L, S and N_b are invariant under the action of this group so that they can be computed explicitly (see [23]).

If $(z,t), (z',t') \in \mathbb{R}^3$ the group operation is given by:

$$(z,t)\cdot(z',t') = (z+z', t+t'+2\,\mathrm{Im}(z\cdot\bar{z}')) . \tag{219}$$

Multiplication by an element of this group can be extended to a holomorphic automorphism of $\bar{\Omega}$ onto $\bar{\Omega}$. The operator S is then given by

$$Sf = f * \frac{1}{\pi^2(|z|-it)^2} \tag{220}$$

where $*$ denotes convolution under the group. The operator $\bar{\partial}_b^* N_b^2 \bar{\partial}_b$ (see (214)) is then given by:

$$\bar{\partial}_b^* N_b^2 \bar{\partial}_b f = f * \frac{1}{2\pi^2(|z|^2-it)} \log\left(\frac{|z|^2-it}{|z|^2+it}\right) . \tag{221}$$

The analysis given in [23] and [24] then prove the desired result.

Finally we remark that the problem of non-solvability for equations with simple characteristics has been completely settled by Nirenberg and Treves (see [25]). In that case non-solvability can only occur if the symbol of the operator is complex. In general however, non-solvability can occur also for operators with real symbol, take for example the operator $P\bar{P}^2P$ where P is given by (215) and $\bar{P} = \frac{\partial}{\partial \bar{z}} - iz\frac{\partial}{\partial t}$.

References

[1] Kohn, J. J. and Nirenburg, L.: "Noncoercive boundary value problems," CPAM, vol. XVIII No. 3, p. 443-492(1965).

[2] Fichera, A., "Sulle equazioni lineari ellittico-paraboliche del secondo ordine," Acc. Naz. Lincei Mem. Ser. 8, Vol. 5, p. 1-30(195).

[3] Kohn, J. J. and Nirenberg, L.: "Degenerate Elliptic-Parabolic Equations of Second Order," CPAM, Vol. XX, 797-872(1967).

[4] Sweeney, W. J., "The D-Neumann problem," Acta Math. 120, 223-277 (1968).

[5] Hörmander, L.: "Hypoelliptic second order differential equations," Acta Math. 119, 147-171(1967).

[6] Kohn, J. J.: "Pseudo-differential operators and hypoellipticity," Proc. Conf. Partial Diff. Eq., pp. 61-99, AMS Pprvidence, R.I. (1971).

[7] Radkevitch, E. V.: "Hypoelliptic operators with multiple characteristics." Mat. Sb. 79(121), 193-216(1969).

[7] Kohn, J. J.: "Pseudo-differential operators and non-elliptic problems," CIME Conf. Stresa 1968 Ed. Cremoneze Rome, 157-165(1969).

[8] Stein, E. and Rothchild, L., "Hypoelliptic differential operators and nilpotent groups," Acta Math. 137, 247-320(1976).

[9] Folland, G. B. and Kohn, J. J.: "The Neumann problem for the Cauchy-Riemann complex," Ann. Math. Studies # 75(1972).

[10] Kohn, J. J.: "Propagation of singularities for the Cauchy-Riemann equations," Proc. C.I.M.E. Conf. on Complex Analysis of June 1973, 179-280.

[11] Hörmander, L., "L_2-Estimates and existence theorems for the $\bar{\partial}$-operator," Acta Math., vol. 113, 89-151(1965).

[12] Kohn, J. J., "Global regularity for $\bar{\partial}$ on weakly pseudo-convex manifolds," Trans. AMS, vol. 181, 273-291(1973).

[13] Kohn, J. J., "Methods of partial differential equations in complex analysis," Proc. of Symp. in Pure Math. vol. XXX part 1; AMS, 213-237(1977).

[14] Kohn, J. J., "Sufficient conditions for subellipticity on weakly pseudo-convex domains," Proc. N.A.S. vol. 74 no. 7, 2214-2216(1977).

[15] Lojasiewicz, S., "Sur le probleme de la division," Studia Math. 87-137(1959).

[16] Diedrich, K. and Fornaess, J. E., "Complex manifolds in real-analytic pseudo-convex hypersurfaces," Proc. N.A.S. (to appear).

[17] Kohn, J. J., "Boundary behaviour of $\bar{\partial}$ on weakly pseudo-convex manifolds of dimension two," J. Diff. Geom. 6, 523-542(1972).

[18] Greiner, P., "On subelliptic estimates of the $\bar{\partial}$-Neumann problem in $\mathbb{C}^2$," J. Diff. Geom. 9, 239-250(1974).

[19] Kohn, J. J. and Rossi, H., "On the extension of holomorphic functions from the boundary of a complex manifold," Ann. of Math., vol. 81 No. 3, 451-472(1965).

[20] Kohn, J. J., "Boundaries of complex manifolds," Proc. Conf. in Complex manifolds, Minneapolis 1964, 81-94.

[21] Trevs, F. J., lectures in this volume.

[22] Lewy, H., "An example of a smooth linear partial differential equation without solution," Ann. Math. 66, 155-158(1957).

[23] Greiner, P., Kohn, J. J. and Stein, E., "Necessary and sufficient conditions for the solvability of the Lewy equation," Proc. N. A. S. vol. 72, No. 9, 3287-3289(1975).

[24] Folland, G. B. and Stein, E., "Estimates for the $\bar{\partial}_b$-complex and analysis on the Heisenberg group," CPAM 27, 429-522(1974).

[25] Nirenberg, L. and Treves, J. F., "On local solvability of linear partial differential equations," Comm. Pure Appl. Math. Part I: vol. 23, 1-38(1970), Part II vol. 23, 459-510(1970).

CENTRO INTERNAZIONALE MATEMATIVO ESTIVO
(C.I.M.E.)

CONDITIONS NECESSAIRES ET SUFFISANTES POUR L'EXISTENCE ET L'UNICITE DES SOLUTIONS DU PROBLEME DE LA DERIVEE OBLIQUE

K. TAIRA

Department of Mathematics, Tokyo - Institute of Technology, Japan - Departement de Mathematiques, Université de Paris-Sud, France

Corso tenuto a Bressanone dal 16 al 24 giugno 1977

Conditions nécessaires et suffisantes pour l'existence et l'unicité des solutions du problème de la dérivée oblique

Kazuaki TAIRA
Department of Mathematics, Tokyo Institute of Technology, Japan
Département de Mathématiques, Université de Paris-Sud, France

Introduction. — Dans cette Note on considère le problème de la dérivée oblique pour le laplacien donnant lieu au principe du maximum et on donne des conditions nécessaires et suffisantes pour l'existence et l'unicité des solutions avec perte d'une dérivée en comparaison du cas coercif dans le cadre des espaces de Sobolev (Théorème 2). Dans la démonstration on utilise des résultats de (8) (cf. (7)) et le théorème 5.9 de Hörmander (4).

1. Formulation du problème. — Soit Ω un domaine borné d'un espace euclidien R^n, $\overline{\Omega}$ étant une variété compacte à bord Γ de classe C^∞, de dimension n.

On considère le problème de la dérivée oblique suivant : Pour deux fonctions f et ϕ définies dans Ω et sur Γ respectivement, trouver une fonction u dans Ω telle que

$$(+)\qquad \begin{cases} \Delta u = f & \text{dans } \Omega, \\ \mathcal{B} u \equiv a \dfrac{\partial u}{\partial \nu} + \alpha u + bu\Big|_{\Gamma} = \phi & \text{sur } \Gamma. \end{cases}$$

Ici Δ est le laplacien sur Ω, a et b sont des fonctions à valeurs réelles et C^∞ sur Γ, ν est la normale unitaire extérieure à Γ, et α est un champ de vecteurs sur Γ.

On étudie le problème de l'existence et l'unicité des solutions de (+) dans le cadre des espaces de Sobolev. Pour établir des théorèmes d'unicité des solutions, on utilise le <u>principe du maximum positif au bord</u> comme dans (8) (cf. (1)). On suppose donc que la condition aux limites $\mathcal{B}$ satisfait au principe du maximum positif au bord :

(PMB) $\quad u \in C^1(\overline{\Omega})$, $x' \in \Gamma$ et $u(x') = \sup u \geqq 0 \Longrightarrow \mathcal{B} u(x') \geqq 0.$

D'après le théorème X de Bony-Courrège-Priouret (1), on voit que la propriété (PMB) est satisfaite si et seulement si

(H.1) $\quad a(x) \geqq 0$ et $b(x) \geqq 0 \quad$ sur Γ.

On obtient alors la

<u>Proposition 1</u>. — <u>On suppose</u> :

(H.1) $\quad a(x) \geqq 0$ <u>et</u> $b(x) \geqq 0 \quad$ <u>sur</u> Γ.

(B.1) $\quad b(x) > 0$ <u>sur</u> $\Gamma_o = \left\{ x \in \Gamma \ : \ a(x) = 0 \right\}$.

<u>On a alors</u> :

$u \in C^2(\Omega) \cap C^1(\overline{\Omega})$, $\Delta u(x) \geqq 0$ <u>dans</u> Ω <u>et</u> $\mathcal{B} u(x) \leqq 0$ <u>sur</u> $\Gamma \Longrightarrow u(x) \leqq 0$ <u>sur</u> $\overline{\Omega}$.

<u>Remarque 1</u>. — Dans le cas où $a(x) > 0$ sur Γ, la proposition 1 reste valable en remplaçant l'hypothèse (B.1) par :

(B.0)' $\quad b(x) \not\equiv 0 \quad$ sur Γ.

2. Réduction au bord. — Pour toute $\varphi \in H^{t-1/2}(\Gamma)$ avec $t \in R$, il existe une solution unique $w \in H^t(\Omega)$ du problème

$$\begin{cases} \Delta w = 0 & \text{dans } \Omega, \\ w|_\Gamma = \varphi & \text{sur } \Gamma. \end{cases}$$

On définit le noyau de Poisson $\mathcal{P} : H^{t-1/2}(\Gamma) \longrightarrow H^t(\Omega)$ par $w = \mathcal{P}\varphi$.

De même, pour toute $f \in H^{s-2}(\Omega)$ avec $s \geqq 2$, il existe une solution unique $v \in H^s(\Omega)$ du problème

$$\begin{cases} \Delta v = f & \text{dans } \Omega, \\ v|_\Gamma = 0 & \text{sur } \Gamma. \end{cases}$$

On définit le noyau de Green $\mathcal{G} : H^{s-2}(\Omega) \longrightarrow H^s(\Omega)$ par $v = \mathcal{G} f$.

On a alors la

Proposition 2. — Pour deux fonctions $f \in H^{s-2}(\Omega)$ et $\phi \in H^{s-3/2}(\Gamma)$ avec $s \geqq 2$, il existe une solution $u \in H^t(\Omega)$ $(t \leqq s)$ du problème (+) si et seulement si il existe une solution $\varphi \in H^{t-1/2}(\Gamma)$ de l'équation

(++) $$T\varphi \equiv \mathcal{B}\mathcal{P}\varphi = \phi - \mathcal{B} v \quad \text{sur } \Gamma,$$

où $v = \mathcal{G} f \in H^s(\Omega)$.

On ramène donc l'étude du problème (+) à celle de l'équation (++) sur Γ. Rappelons que l'opérateur $T \equiv \mathcal{B}\mathcal{P}$:

$$\varphi \longrightarrow \mathcal{B}\mathcal{P}\varphi = a \frac{\partial}{\partial \nu}(\mathcal{P}\varphi)\Big|_\Gamma + \alpha\varphi + b\varphi$$

est un opérateur pseudo-différentiel du premier ordre sur Γ (cf. (3)).

3. Étude de T. — Pour calculer le symbole de l'opérateur pseudo-différentiel T , on a besoin du

Lemme 1. — Posons :

$$\Pi\varphi = \frac{\partial}{\partial \nu}(\mathcal{P}\varphi)\Big|_\Gamma .$$

Alors $\Pi = \Pi^*$, *où* Π^* *est l'adjoint formel de* Π.

Ce lemme se démontre en utilisant la formule de Green.

Soit $(x,\xi) = (x_1, x_2, \ldots, x_{n-1}, \xi_1, \xi_2, \ldots, \xi_{n-1})$ des coordonnées locales du fibré cotangent $T^*\Gamma$ et $(g_{ij}(x))_{1 \leqq i,j \leqq n-1}$ la métrique riemannienne de Γ induite par la métrique naturelle de R^n. On désigne par $\omega_x(\ ,\ )$ la deuxième forme fondamentale en un point x de l'hypersurface $\Gamma \subset R^n$ et par $M(x)$ la courbure moyenne en un point x de Γ respectivement.

En utilisant un résultat de Fujiwara-Uchiyama (2) et le lemme 1, on peut calculer le symbole de T.

Lemme 2. — Le symbole de l'opérateur pseudo-différentiel $T = a\Pi + \alpha + b$ *est donné par :*

$$\left[a(x)|\xi| + \sqrt{-1}\sum_{k=1}^{n-1}\alpha^k(x)\xi_k\right] + \left[b(x) + \frac{1}{2}a(x)\left(\frac{\omega_x(\hat{\xi},\hat{\xi})}{|\xi|^2} - (n-1)M(x)\right.\right.$$

$$\left.\left. + \sqrt{-1}\,a(x)\left(\sqrt{\rho(x)}\sum_{j=1}^{n-1}\frac{\partial}{\partial\xi_j}(|\xi|)\frac{\partial}{\partial x_j}\left(\frac{1}{\sqrt{\rho(x)}}\right) - \frac{1}{2}\sum_{j=1}^{n-1}\frac{\partial^2}{\partial x_j\,\partial\xi_j}(|\xi|)\right)\right]$$

+ *des termes d'ordre* $\leqq -1$.

Ici $|\xi| = \sqrt{g^{ij}(x)\xi_i\xi_j}$, $\rho(x) = \left(\det(g_{ij}(x))\right)^{1/2}$, α^k est une composante de α et $\hat{\xi} = (g^{1j}(x)\xi_j, g^{2j}(x)\xi_j, \ldots, g^{n-1\,j}(x)\xi_j)$, où $(g^{ij}(x))$ est la matrice inverse de $(g_{ij}(x))$.

D'après le lemme 2, on voit que l'opérateur T est *elliptique* si et seulement si la fonction a ne s'annule en aucun point de Γ, i.e., si et seulement si

(A.0) $\quad a(x) > 0 \quad$ sur Γ.

En utilisant une variante d'une méthode de Agmon-Nirenberg comme dans (8), on obtient la

Proposition 3. — On suppose :

(A.0) $a(x) > 0$ sur Γ.

Alors, pour tout $s \geqq 3/2$, l'opérateur $T : H^{s-1/2}(\Gamma) \longrightarrow H^{s-3/2}(\Gamma)$ est d'indice zéro.

En remarquant que $\{\varphi \in H^{s-1/2}(\Gamma) \;:\; T\varphi = 0\} \subset C^{\infty}(\Gamma)$, on obtient d'après la remarque 1 et la proposition 2 le

Corollaire 1. — On suppose :

(H.1) $a(x) \geqq 0$ et $b(x) \geqq 0$ sur Γ.

Alors, pour tout $s \geqq 3/2$, l'opérateur T est un isomorphisme algébrique et topologique de $H^{s-1/2}(\Gamma)$ sur $H^{s-3/2}(\Gamma)$ si et seulement si

(A.0) $a(x) > 0$ sur Γ.

(B.0)' $b(x) \not\equiv 0$ sur Γ.

Pour étudier des conditions nécessaires et suffisantes pour que, pour tout $s \geqq 3/2$, l'opérateur $T : H^{s-3/2}(\Gamma) \longrightarrow H^{s-3/2}(\Gamma)$ soit un isomorphisme algébrique avec perte d'une dérivée en comparaison du cas coercif, i.e., du cas où $a(x) > 0$ sur Γ, introduisons un opérateur linéaire non borné et fermé $\mathcal{T} : H^{s-3/2}(\Gamma) \longrightarrow H^{s-3/2}(\Gamma)$ de la manière suivante :

a) Le domaine de définition de $\mathcal{T}$ est $\mathcal{D}(\mathcal{T}) = \{\varphi \in H^{s-3/2}(\Gamma) : T\varphi \in H^{s-3/2}(\Gamma)\}$.

b) $\mathcal{T}\varphi = T\varphi$ pour toute $\varphi \in \mathcal{D}(\mathcal{T})$.

En raisonnant comme dans la démonstration du corollaire 4.4 de (6), on déduit du lemme 2 et du théorème 3.1 de Melin (5) la

Proposition 4. — Soit $s \geqq 3/2$. On suppose :

(A.1) $a(x) \geqq 0$ sur Γ.

$(C)_s$ En tout point $x \in \Gamma_0 = \{x \in \Gamma : a(x) = 0\}$,

$$2b(x) - \operatorname{div} \alpha(x) + (s - 3/2)\left\{|\xi|^2, \alpha^k(x)\,\xi_k\right\} > 0$$

pour tout $\xi \in T_x^*\Gamma$ avec $0 \leqq |\xi| \leqq 1$.

<u>Alors, l'opérateur</u> $\mathcal{T} : H^{s-3/2}(\Gamma) \longrightarrow H^{s-3/2}(\Gamma)$ <u>est d'indice zéro</u>.

Ici div α est la divergence de α par rapport à $(g_{ij}(x))$ et

$$\{f,g\} = \sum_{j=1}^{n-1}\left(\frac{\partial f}{\partial \xi_j}\frac{\partial g}{\partial x_j} - \frac{\partial f}{\partial x_j}\frac{\partial g}{\partial \xi_j}\right)$$

est la crochet de Poisson de f et g.

D'après le lemme de Sobolev, la proposition 1 et la proposition 2, on a le

<u>Corollaire 2</u>. — <u>Soit</u> $s \geqq [n/2] + 4$. <u>On suppose</u> :

(H.1) $a(x) \geqq 0$ <u>et</u> $b(x) \geqq 0$ <u>sur</u> Γ.

(B.1) $b(x) > 0$ <u>sur</u> $\Gamma_0 = \{x \in \Gamma : a(x) = 0\}$.

$(C)_s$ <u>En tout point</u> $x \in \Gamma_0$,

$$2b(x) - \operatorname{div} \alpha(x) + (s - 3/2)\left\{|\xi|^2, \alpha^k(x)\xi_k\right\} > 0$$

<u>pour tout</u> $\xi \in T_x^*\Gamma$ <u>avec</u> $0 \leqq |\xi| \leqq 1$.

<u>Alors, l'opérateur</u> $\mathcal{T} : H^{s-3/2}(\Gamma) \longrightarrow H^{s-3/2}(\Gamma)$ <u>est un isomorphisme algébrique</u>.

De plus, en utilisant le théorème 5.9 de (4), on peut démontrer le

<u>Corollaire 3</u>. — <u>On suppose</u> :

(H.1) $a(x) \geqq 0$ <u>et</u> $b(x) \geqq 0$ <u>sur</u> Γ.

(H.2) <u>Il existe une constante</u> $C_0 > 0$ <u>telle que</u>

$$\left|\sum_{k=1}^{n-1} \alpha^k(x)\xi_k\right| \leqq C_0\, a(x)\,|\xi| \qquad \text{sur } T^*\Gamma.$$

<u>Alors, pour tout</u> $s \geqq 3/2$, <u>l'opérateur</u> $\mathcal{T} : H^{s-3/2}(\Gamma) \longrightarrow H^{s-3/2}(\Gamma)$ <u>est un isomorphisme algébrique si et seulement si</u>

(B.1) $b(x) > 0$ <u>sur</u> $\Gamma_0 = \{x \in \Gamma : a(x) = 0\}$.

4. <u>Conditions nécessaires et suffisantes</u>. — En utilisant la proposition 2 avec $t = s$ et le corollaire 1, on obtient d'abord le

Théorème 0. — Soit $s \geqq 2$. On suppose :

(H.1) $a(x) \geqq 0$ et $b(x) \geqq 0$ sur Γ.

Alors, le problème (+) admet une solution unique u dans $H^s(\Omega)$ pour toutes $f \in H^{s-2}(\Omega)$ et $\phi \in H^{s-3/2}(\Gamma)$ si et seulement si les hypothèses (A.0) et (B.0)' sont satisfaites :

(A.0) $a(x) > 0$ sur Γ.

(B.0)' $b(x) \not\equiv 0$ sur Γ.

De même, on obtient d'après la proposition 2 avec $t = s-1$ et le corollaire 2 le

Théorème 1. — Soit $s \geqq [n/2] + 4$. On suppose :

(H.1) $a(x) \geqq 0$ et $b(x) \geqq 0$ sur Γ.

(B.1) $b(x) > 0$ sur $\Gamma_0 = \{x \in \Gamma : a(x) = 0\}$.

$(C)_s$ En tout point $x \in \Gamma_0$,

$$2b(x) - \operatorname{div} \alpha(x) + (s - 3/2) \{|\xi|^2, \alpha^k \xi_k\} > 0$$

pour tout $\xi \in T_x^* \Gamma$ avec $0 \leqq |\xi| \leqq 1$.

Alors, le problème (+) admet une solution unique u dans $H^{s-1}(\Omega)$ pour toutes $f \in H^{s-2}(\Omega)$ et $\phi \in H^{s-3/2}(\Gamma)$.

Remarque 2. — Si l'hypothèse (C.1) dans le théorème 1 de (8) :

(C.1) α s'annule au second ordre sur Γ_0

est satisfaite, l'hypothèse $(C)_s$ se réduit à l'hypothèse (B.1), donc le théorème 1 ci-dessus généralise le théorème 1 de (8).

De plus, d'après le corollaire 3, on obtient le

Théorème 2. — Soit $s \geqq 2$. On suppose :

(H.1) $a(x) \geqq 0$ et $b(x) \geqq 0$ sur Γ.

(H.2) Il existe une constante $C_0 > 0$ telle que

$$\left|\sum_{k=1}^{n-1} \alpha^k(x) \xi_k\right| \leqq C_0 \, a(x) \, |\xi| \quad \text{sur } T^*\Gamma.$$

Alors, le problème (+) admet une solution unique u dans $H^{s-1}(\Omega)$ pour toutes $f \in H^{s-2}(\Omega)$ et $\phi \in H^{s-3/2}(\Gamma)$ si et seulement si l'hypothèse (B.1) est satisfaite :

$$\text{(B.1)} \qquad b(x) > 0 \qquad \underline{\text{sur}} \qquad \Gamma_0 = \left\{ x \in \Gamma \ : \ a(x) = 0 \right\}.$$

Bibliographie

(1) J.M. Bony, P. Courrège et P. Priouret, Semi-groupes de Feller sur une variété à bord compacte et problèmes aux limites intégro-différentiels du second ordre donnant lieu au principe du maximum (Ann. Inst. Fourier vol. 18, 1968, p. 369-521).

(2) D. Fujiwara and K. Uchiyama, On some dissipative boundary value problems for the Laplacian (J. Math. Soc. Japan, vol. 27, 1971, p. 625-635).

(3) L. Hörmander, Pseudo-differential operators and non-elliptic boundary problems (Ann. of Math., vol. 83, 1966, p. 129-209).

(4) L. Hörmander, A class of hypoelliptic pseudodifferential operators with double characteristics (Math. Ann., vol. 217, 1975, p. 165-188).

(5) A. Melin, Lower bounds for pseudo-differential operators (Ark. för Mat., vol. 9, 1971, p. 117-140).

(6) K. Taira, On some non-coercive boundary value problems for the Laplacian (J. Fac. Sci. Univ. Tokyo, Sec. IA, vol. 23, p. 343-367).

(7) K. Taira, Sur le problème de la dérivée oblique (C. R. Acad. Sc. Paris, t. 284, série A, 1977, p. 1511-1513).

(8) K. Taira, Sur le problème de la dérivée oblique (à paraître).

CENTRO INTERNAZIONALE MATEMATICO ESTIVO

(C.I.M.E.)

BOUNDARY VALUE PROBLEMS FOR ELLIPTIC EQUATIONS

F. TREVES

Rutgers University, U.S.A.

Corso tenuto a Bressanone dal 16 al 24 giugno 1977

BOUNDARY VALUE PROBLEMS FOR ELLIPTIC EQUATIONS

François Treves

Rutgers University, U. S. A.

Introduction

This series of lectures presents a systematic treatment of boundary value problems for elliptic equations - without a priori distinguishing between the coercive problems (also called of the Lopatinski-Shapiro type) and the noncoercive ones. The property we are concerned with here is that of regularity up to the boundary. Now new results will be proved: what is shown is that known results can be "integrated" in a unified approach. Nor is the approach very novel: it is essentially that of transfering the problem to the boundary, where it turns into an interior problem. In one version or another it has been much used in the past, perhaps first by Calderon, shortly afterwards by Seeley, Hörmander, Vishik, etc. If there could be any claim of novelty in the present treatment (and the author is not even sure of that) it would lie in the manner in which the transfer to the boundary is effected: In the neighborhood of the boundary the elliptic equation (or system) $Pw = f$ is decomposed into a couple of equations

(1) $\quad \frac{\partial u}{\partial t} - A^{-}(t)u = g, \qquad$ (2) $\quad \frac{\partial v}{\partial t} - A^{+}(t)v = h ,$

where $A^{+}(t)$ (resp., $A^{-}(t)$) is a pseudodifferential operator on the boundary, depending smoothly on the variable t transversal to the boundary, elliptic negative (resp., elliptic positive) of order one (in the case we discuss; it could be of different order in other cases). As for g and h they are distributions on the boundary, depending on t; g is a functional of v, usually $g = Jv$, where J is a simple operator. The boundary conditions bearing on w translate into a single boundary condition bearing on the solution of (1):

$$(3) \qquad \mathcal{B}\, u(0) = u_0 ,$$

with u_0 a distribution on the boundary, which may also depend on the solution v of (2) (and, of course, on the data). Then one solves Eq. (2) "backward" , that is, from a certain value T of t which defines a hypersurface parallel to the boundary inside the domain, to $t = 0$ (which defines the boundary), and after this, one puts the value of v in (1) and (3), and solves this Cauchy problem forward, with initial datum u(0).

The main advantage of such an approach is that we get all we want without introducing any tool more sophisticated than standard pseudodifferential operators on the boundary, those called of type (1,0) by Hörmander (see [6]). Indeed not only is $\mathcal{B}$ of that type (in most applications it will be even better: it will be a classical pseudodifferential operator, that is, an asymptotic series of operators with homogeneous symbols) but so will be the parametrices of the forward Cauchy problem for (1) and of the backward Cauchy problem for (2). In particular they decrease the wave-front sets (of g, h, u(0)). Thus everything is reduced to deducing the regularity of the initial value u(0) from that of the datum u_0 in (3). This reduction is established in Chapters II, III.

In Chapter IV we show that the problems of Lopatinski-Shapiro type are characterized by the property that $\mathcal{B}$ is elliptic (this had already been pointed out by Calderon; see [3]). Thus the theorems of regularity up to the boundary ([1], [9]) reduce to the assertion that pseudodifferential operators are pseudolocal, and act in a certain manner on the Sobolev spaces (by using the properties of the parametrices of (1) & (2) we derive the coercive estimates). Although this will not be discussed here the same observation applies to analytic regularity: the parametrices of (1) & (2) also decrease the analytic wave-front sets (keep in mind that $A^{\pm}(t)$ have order one!).

In Ch. V we describe the $\bar{\partial}$-Neumann problem and determine, to a certain extent, the corresponding boundary operator $\mathcal{B}$: we determine its principal symbol, and also its subprincipal symbol at the points of its characteristic set (which is a smooth conic manifold and on which the principal symbol vanishes exactly of order two). Availing ourselves of this information we show how the known necessary and sufficient conditions for the validity of the subelliptic $\frac{1}{2}$-estimate are a particular case of recent results on certain classes of pseudodifferential operators with double characteristics (see [2], [7]).

In Ch. VI we discuss rapidly the boundary value problems we call of principal type: $\mathcal{B}$ is then a pseudodifferential operator of principal type (Def. VI.1). Here as in the preceding chapters we do not prove any "theorem" : we state two results, one of the author essentially characterizing the hypo-ellipticity of differential operators of principal type, the other one of Yu. Egorov, characterizing the subelliptic pseudodifferential operators.

Ch. I gives a quick overview, without proofs, of pseudodifferential operator theory. One of the motivations for such a survey is to "fix" the terminology and later avoid confusion. It is important that we know what we are dealing with, be it true operators or classes of such modulo regularizing ones. It should be admitted, at this point, that the decomposition (1)-(2)-(3) of the original boundary value problem is only valid modulo regularizing operators, and so is the entire argument in these lectures. But because of our special point of view - because we only look at regularity theorems - this does not endanger our conclusions.

CONTENTS

I. BACKGROUND ON PSEUDODIFFERENTIAL OPERATORS. WAVE-FRONT SETS

We shall systematically use standard notation: If $x_1, \ldots, x_n$ are coordinates in Euclidean space $\mathbb{R}^n$, or local coordinates in a chart of some (always C^∞, i. e., smooth) manifold X, by ∂_x^α we mean the monomial $(\partial/\partial x_1)^{\alpha_1} \ldots (\partial/\partial x_n)^{\alpha_n}$, by D_x^α we mean $i^{-|\alpha|}\partial_x^\alpha$. If E is a locally convex topological vector space, by $C_c^\infty(X;E)$ we denote the space of compactly supported C^∞ functions in X with values in E, by $\mathcal{D}'(X;E)$ the space of E-valued distributions in X, by $\mathcal{E}'(X;E)$ the subspace of $\mathcal{D}'(X;E)$ consisting of the compactly supported distributions. All these spaces, and all other standard spaces we might use (such as $C^\infty(X;E)$, the space of C^∞ mappings $X \to E$) are endowed with their standard locally convex topologies. Whenever $E = \mathbb{C}$, the complex plane, we shall omit mentioning it and write, e. g., $C_c^\infty(X)$ instead of $C_c^\infty(X;\mathbb{C})$, $\mathcal{D}'(X)$ instead of $\mathcal{D}'(X;\mathbb{C})$, etc. In practice E will almost always be a finite dimensional vector space over the complex numbers. But sometimes it will be a Hilbert space or the Banach space of bounded linear operators acting on it (if H is the Hilbert space, we denote by L(H) the space of continuous endomorphisms of H).

Let X, Y be two smooth manifolds, E, F two locally convex spaces. We recall the Schwartz kernels theorem: any continuous linear map of $C_c^\infty(X;E)$ into $\mathcal{D}'(Y;F)$ is of the form

$$f(x) \mapsto \int K(x,y)f(x)dx \,, \tag{I.1}$$

where $K(x,y)$ is a distribution in $X \times Y$ valued in $L(E;F)$, the space of continuous linear operators $E \to F$ suitably topologized (E and F must be "reasonable" spaces, which is definitely the case for Banach spaces, Frechet spaces, spaces like $\mathcal{D}'$, $\mathcal{E}'$, C_c^∞, etc.; also the distinction between distributions as currents either of degree zero or of maximum degree should be maintained, but we shall disregard it here.)

Let us denote by K the continuous linear mapping (I.1). We say that the kernel K(x,y) is separately regular in x and in y if K defines a continuous linear of $C_c^\infty(X;E)$ into $C^\infty(Y;F)$ and if K extends as a continuous linear map of $\mathcal{E}'(X;E)$ into $\mathcal{D}'(Y;F)$. Interms of completed topological tensor products this is equivalent with the property that

$$\text{(I.2)} \qquad K(x,y) \in \left\{ \left[\mathcal{D}'(X) \hat{\otimes} C^\infty(Y) \right] \cap \left[C^\infty(X) \hat{\otimes} \mathcal{D}'(Y) \right] \right\} \hat{\otimes} L(E;F) .$$

If E and F are both reflexive, the separate regularity of K(x,y) is equivalent with the fact that K maps $C_c^\infty(X;E)$ into $C^\infty(Y;F)$ and its transpose, ${}^tK$, maps (linearly and continuously) $C_c^\infty(Y;F')$ into $C^\infty(X;E')$ (E', F' are the respective topological duals of E and F). Or with the fact that K extends as a continuous linear map $\mathcal{E}'(X;E) \to \mathcal{D}'(Y;F)$ and that its transpose ${}^tK$ extends as a continuous linear map $\mathcal{E}'(Y;F') \to \mathcal{D}'(X;E')$.

Suppose momentarily that Y = X. The mapping (I.1) is said to be very regular if it is separately regular in x and in y and if, moreover, the kernel $K(x,y) \in \mathcal{D}'(X \times X;L(E;F))$ is a C^∞ function in the complement of the diagonal of $X \times X$ (valued in L(E;F)).

Let us go back to the general case, with Y not necessarily equal to X. We shall say that the operator (I.1) is regularizing if it extends as a continuous linear map of $\mathcal{E}'(X;E)$ into $C^\infty(Y;F)$. In order that this be the case it is necessary and sufficient that $K(x,y)$ be a C^{∞} function in $X \times Y$ with values in $L(E;F)$. Notice that, then, the transpose ${}^tK$ is a continuous linear map of $\mathcal{E}'(Y;F')$ into $C^{\infty}(X;E')$.

The last of this list of definitions concerning mappings (I.1) is that of properly supported: the kernel $K(x,y)$ (or the mapping K) is said to be properly supported if K maps $C_c^{\infty}(X;E)$ into $\mathcal{E}'(Y;F)$ and if, moreover, its transpose ${}^tK$ maps $C_c^{\infty}(Y;F')$ into $\mathcal{E}'(X;E')$. This is equivalent with saying that the kernel $K(x,y)$ has the following property:

(I.3) Given any compact subset $\mathcal{K}$ of X there is a compact subset $\mathcal{K}'$ of Y such that, for any $\phi \in C^{\infty}(X)$ with $\operatorname{supp} \phi \subset \mathcal{K}$ ($\operatorname{supp} \phi$ = support of ϕ), we have $\operatorname{supp}(\phi(x)K(x,y)) \subset \mathcal{K} \times \mathcal{K}'$; and given any compact subset $\mathcal{K}_1'$ of Y there is a compact subset $\mathcal{K}_1$ of X such that, for any $\psi \in C^{\infty}(Y)$ with $\operatorname{supp} \psi \subset \mathcal{K}_1'$, the support of $\psi(y)K(x,y)$ lies in $\mathcal{K}_1 \times \mathcal{K}_1'$.

If $K(x,y)$ is properly supported, K extends as a continuous linear map $C^{\infty}(X;E) \to \mathcal{D}'(Y;F)$, and ${}^tK$ as one from $C^{\infty}(Y;F')$ into $\mathcal{D}'(X;E')$. If we combine this with the definition of separately regular operators we see that

(I.4) If $K(x,y)$ is properly supported and separately regular in x and y, K defines a continuous linear map of $C_c^{\infty}(X;E)$ (resp., $C^{\infty}(X;E)$, resp., $\mathcal{E}'(X;E)$, resp., $\mathcal{D}'(X;E)$) into $C_c^{\infty}(Y;F)$ (resp., $C^{\infty}(Y;F)$, resp., $\mathcal{E}'(Y;F)$, resp., $\mathcal{D}'(Y;F)$) .

We now recall some properties of very regular kernels. Thus we assume, throughout the remainder, that $X = Y$.

The <u>singular support</u> of a distribution u, that is to say, the smallest closed subset in the complement of which u is a C^∞ functions, will be denoted by sing supp u. The following result is classical (cf. [12], Th. 52.1):

<u>Theorem I.1.- Suppose that the kernel</u> $K(x,y)$ <u>is very regular. Then</u>

(I.5) $$\text{sing supp } Ku \subset \text{sing supp } u\ , \quad \forall\, u \in \mathcal{E}'(X;E)$$

(<u>i. e.</u>, Ku is an F-valued C^∞ function in every open subset of X in which u is an E-valued C^∞ function).

Th. I.1 is often restated by saying that a very regular operator K is <u>pseudodlocal</u>. A standard application is that, if there is a very regular, properly supported linear operator $P : \mathcal{E}'(X;F) \to \mathcal{E}'(X;E)$ such that KP = Identity of $\mathcal{E}'(X;F)$, then

$$\text{sing supp } Pu = \text{sing supp } u, \qquad \forall\, u \in \mathcal{E}'(X;F).$$

Another property of very regular kernels is the following:

(I.6) <u>Let</u> $K(x,y)$ <u>be a very regular kernel in</u> $X\times X$, <u>valued in</u> $L(E;F)$. <u>There is a properly supported kernel</u> $K_1(x,y) \in \mathcal{D}'(X\times X;L(E;F))$ <u>such that</u> $K - K_1 \in C^\infty(X\times X;L(E;F))$.

<u>Proof</u>: $K_1(x,y) = \phi(x,y)K(x,y)$ with $\phi \in C^\infty(X\times X)$ properly supported (cf. (I.3)) and equal to one ine a neighborhood of the diagonal.

Pseudodifferential operators are a special case of very regular operators of the kind (I.1). We now recall their definitions and main properties. First we define pseudodifferential operators in an open subset U of a Euclidean space R^n (where the variable is $x = (x_1,\dots,x_n)$; the dual coordinates are denoted by $\xi_1,\dots,\xi_n$ and we write $x\cdot\xi = x_1\xi_1+\dots+x_n\xi_n$). For the moment we restrict ourselves to distributions and functions with complex values; the case of more general values, say lying in a Banach space, is gotten by straightforward generalization). We shall systematically use the Fourier transform,

(I.7) $$\hat{u}(\xi) = \int e^{-ix\cdot\xi}\, u(x)\, dx,$$

and the Fourier inversion formula,

(I.8) $$u(x) = (2\pi)^{-n}\int e^{ix\cdot\xi}\,\hat{u}(\xi)\, d\xi\ ,$$

say for u in $\mathcal{S}(\mathbb{R}^n)$ (or generalized in the various standard manners).

For any real m we denote by $S^m(U,U)$ the space of C^∞ functions $a(x,y,\xi)$ in $U\times U\times\mathbb{R}^n$ having the following property:

(I.9) To every compact subset $\mathcal{K}$ of $U\times U$, to every triple of elements α, β, γ of $\mathbb{Z}^n_+$ there is $C = C(\mathcal{K};\alpha,\beta,\gamma) > 0$ such that

(I.10) $$|\partial_x^\alpha\partial_y^\beta\partial_\xi^\gamma a(x,y,\xi)| \leq C(1+|\xi|)^{m-|\gamma|}\ ,\ (x,y)\in\mathcal{K},\ \xi\in\mathbb{R}^n.$$

(More generally, the space $S^m_{\rho,\delta}(U,U)$ is defined by substituting $m+\delta|\alpha+\beta| - \rho|\gamma|$ for $m-|\gamma|$ in the exponent at the right, in (I.10); usually one takes $0\leq\delta\leq\frac{1}{2}\leq\rho\leq 1$.) The space $S^m(U,U)$ is topologized in the obvious manner (it is then a Frechet space).

Since $a(x,y,\xi)\in S^m(U,U)$ is tempered with respect to ξ, we may form its Fourier transform, $\hat{a}(x,y,z)$. We introduce then the kernel-distribution $A(x,y) = (2\pi)^{-n}\hat{a}(x,y,y-x)$, that is,

(I.11) $$A(x,y) = (2\pi)^{-n}\int e^{i(x-y)\cdot\xi}\, a(x,y,\xi)\, d\xi\ .$$

We denote by $OPS^m(U,U)$ the space of linear operators defined by kernels of the kind (I.11) : If u is any element of $C_c^\infty(U)$,

(I.12) $$Au(x) = \int A(x,y)u(y)dy = (2\pi)^{-n}\iint e^{i(x-y)\cdot\xi}\, a(x,y,\xi)u(y)dyd\xi\ .$$

Notice that if $a(x,y,\xi) = a(x,\xi)$ is independent of y, (I.12) reads

(I.13) $$Au(x) = (2\pi)^{-n}\int e^{ix\cdot\xi}\, a(x,\xi)\hat{u}(\xi)\, d\xi\ .$$

Definition I.1.- We say that a continuous linear operator $C_c^\infty(U) \to \mathcal{D}'(U)$ is a pseudodifferential operator of order m in U if there is an operator $A \in OPS^m(U,U)$ such that P - A is regularizing (in U). The space of pseudodifferential operators of order m in U will be denoted by $\Psi^m(U)$.

The union of the $\Psi^m(U)$, $m \in \mathbb{R}$, will be denoted by $\Psi(U)$; their intersection by $\Psi^{-\infty}(U)$ (its elements are the regularizing operators in U).

The continuity properties of pseudodifferential operators are best described by means of the Sobolev spaces: $H^s = H^s(\mathbb{R}^n)$ is the space of tempered distributions u in $\mathbb{R}^n$ such that

$$\|u\|_s = \left\{ (2\pi)^{-n} \int |\hat{u}(\xi)|^2 (1 + |\xi|^2)^s \, d\xi \right\}^{\frac{1}{2}} < +\infty ; \tag{I.14}$$

s can be any real number. We recall that $\| \ \|_s$ is a Hilbert norm on H^s, which thus is a Hilbert space, actually a copy of $H^0 = L^2(\mathbb{R}^n)$ for the obvious isomorphism $u \mapsto (1 - \Delta_x)^{s/2} u$ (isomorphism from H^s onto L^2:

$$(1 - \Delta_x)^{s/2} u(x) = (2\pi)^{-n} \int e^{ix\cdot\xi} (1 + |\xi|^2)^{s/2} \hat{u}(\xi) \, d\xi \ .) \tag{I.15}$$

We have the injections with dense images $\mathcal{S} \hookrightarrow H^s \hookrightarrow \mathcal{S}'$. By transposing the first one we obtain the natural injection of the dual $(H^s)'$ of H^s into $\mathcal{S}'$, whose image is exactly H^{-s} (thus H^s and H^{-s} are dual of one another, and $(1 - \Delta_x)^s$ is the canonical isometry of H^s onto its <u>antidual</u>, H^{-s}). If $s \geq s'$, H^s is contained and dense in $H^{s'}$, and the norm $\| \ \|_s$ is larger than the norm $\| \ \|_{s'}$ on H^s. Etc. (cf. [13], Ch. 24 & 25).

Now, if U is any open subset of $\mathbb{R}^n$, by $H^s_{loc}(U)$ one denotes the space of distributions in U such that $\phi u \in H^s$ for every $\phi \in C_c^\infty(U)$, by $H^s_c(U)$ the space of elements of H^s compactly supported in U. These spaces can be equipped with natural locally convex topologies (then $H^s_{loc}(U)$ is a reflexive Fréchet space); $H^s_{loc}(U)$ and $H^{-s}_c(U)$ can be identified to the dual of each other. The following relations are true setwise (the first one is also true topologically):

$$(I.16)\qquad \begin{aligned} C^{\infty}(U) &= \bigcap_s H^s_{loc}(U), &\qquad \mathcal{D}'^{F}(U) &= \bigcup_s H^s_{loc}(U), \\ C^{\infty}_c(U) &= \bigcap_s H^s_c(U), &\qquad \mathcal{E}'(U) &= \bigcup_s H^s_c(U) \end{aligned}$$

($\mathcal{D}'^{F}$ means "distributions of finite order".) The proof of the results we now state can be found in many a text on pseudodifferential operator theory (see e. g.

<u>Theorem I.2.- Any operator</u> $P \in \Psi^m(U)$ <u>induces a continuous linear map</u> $H^s_c(U) \to H^{s-m}_{loc}(U)$, <u>whatever the real number</u> s.

From Th. I.2 and (I.16) one derives

<u>Corollary I.1.-</u> $P \in \Psi(U)$ <u>induces a continuous linear map</u> $C^{\infty}_c(U) \to C^{\infty}(U)$, $\mathcal{E}'(U) \to \mathcal{D}'(U)$ (in other words, the kernel distribution associated with P is separately regular).

<u>Theorem I.3.- Every pseudodifferential operator in</u> U <u>is pseudolocal</u> (its associated kernel is very regular).

<u>Theorem I.4.- If</u> $P \in \Psi^m(U)$ <u>the transpose</u> ${}^tP$ <u>of</u> P <u>and its adjoint</u>, $P^* = {}^t\overline{P}$, <u>also belong to</u> $\Psi^m(U)$.

<u>If</u> Q <u>is an operator belonging to</u> $\Psi^{m'}(U)$ <u>and is properly supported</u>, <u>the compose</u> PoQ <u>belongs to</u> $\Psi^{m+m'}(U)$.

The requirement that Q be properly supported insures that the compose PoQ makes sense.

If we want to define pseudodifferential operators on a manifold we need to know that the concept of pseudodifferential operators in open subsets of Euclidean space is invariant under coordinates changes. Thus let V denote another open subset of $\mathbb{R}^n$, ϕ a diffeomorphism of U onto V, and consider the diagram

$$
\text{(I.17)}\qquad
\begin{array}{ccc}
\mathcal{E}'(U) & \xrightarrow{P} & \mathcal{D}'(U) \\
\phi_* \downarrow & & \downarrow \phi_* \\
\mathcal{E}'(V) & \xrightarrow{P^\phi} & \mathcal{D}'(V)
\end{array} .
$$

In (I.17) P^ϕ is defined as the compose $\phi_* \circ P \circ \phi_*^{-1}$ (ϕ_* is the direct image map on distributions from U to V, ϕ_*^{-1} its inverse).

Theorem I.5.- If $P \in \Psi^m(U)$ then $P^\phi \in \Psi^m(V)$.

If we do not require that V, in (I.17), be an open subset of $\mathbb{R}^n$, but only that it be a C^∞ manifold diffeomorphic to such a subset, we may use that diagram to define P^ϕ as a pseudodifferential operator (of order m) in the manifold V. It is then easy to extend the definition to an arbitrary (smooth) manifold X :

Let P be a continuous linear operator $\mathcal{E}'(X) \to \mathcal{D}'(X)$. If V is any open subset of X, we denote by P_V the composition

$$
\text{(I.18)}\qquad \mathcal{E}'(V) \to \mathcal{E}'(X) \xrightarrow{P} \mathcal{D}'(X) \to \mathcal{D}'(V) ,
$$

where the first arrow stands for the natural injection mapping, the third one for the natural restriction mapping (of distributions from X to V).

Definition I.2.- We say that the operator P is a pseudodifferential operator (resp., of order m) in X if given any open subset V of X diffeomorphic to an open subset of a Euclidean space, the "induced" operator P_V is a pseudodifferential operator (resp., of order m) in V.

We shall denote by $\Psi^m(X)$ the space of pseudodifferential operators of order m in X, by $\Psi(X)$ (resp., $\Psi^{-\infty}(X)$) the union (resp., the intersection) of the spaces $\Psi^m(X)$, $m \in \mathbb{R}$. It follows from Th. I.2 that $\Psi^{-\infty}(X)$ is the space of regularizing operators in X.

The composition of two elements P, Q of $\Psi(X)$, PoQ, might not be defined unless we specify that Q is properly supported (cf. Th. I.5). This restriction is removed if we deal with <u>equivalence classes of pseudodifferential operators modulo regularizing ones</u>, <u>i</u>. <u>e</u>., with elements of the quotient spaces

$$\dot{\Psi}(X) = \Psi(X)/\Psi^{-\infty}(X) \quad (\text{or of } \dot{\Psi}^m(X) = \Psi^m(X)/\Psi^{-\infty}(X)). \tag{I.19}$$

Indeed, by (I.6), we know that every class in $\dot{\Psi}(X)$ contains a representative which is properly supported. Thus, if $\dot{P}$, $\dot{Q} \in \dot{\Psi}(X)$, their compose $\dot{P}o\dot{Q}$ is defined as the class of an operator PoQ where $P \in \dot{P}$, $Q \in \dot{Q}$ and Q is properly supported.

Another advantage in dealing with classes mod $\Psi^{-\infty}(X)$ is that we can make them act on arbitrary distributions, whether they have compact support or not, <u>provided we are willing to reason modulo</u> C^∞ : If $\dot{P} \in \dot{\Psi}(X)$ and if $P \in \dot{P}$ is properly supported, we may define $\dot{P}\dot{u}$ for any element $\dot{u}$ of $\mathcal{D}'(X)/C^\infty(X)$ as the class of Pu, where u is any representative of $\dot{u}$ (recalling (I.4)).

In practice, often without recalling it, one deals with equivalence classes of pseudodifferential operators modulo regularizing ones, rather than with the pseudodifferential operators themselves.

We come now to the <u>symbolic</u> <u>calculus</u> of pseudodifferential operators. For this we go back to the open set U in Euclidean space $\mathbb{R}^n$. Let us refer to the elements of $S^m(U,U)$ as <u>amplitudes</u>. By a <u>symbol</u> <u>of</u> <u>degree</u> m we then mean an amplitude belonging to $S^m(U,U)$ $a(x,y,\xi) = a(x,\xi)$ <u>independent of the second variable</u>, y. These symbols form a subspace $S^m(U)$ of $S^m(U,U)$, topologized in an obvious manner. (Similarly one can define the symbol spaces $S^m_{\rho,\delta}(U)$.)

At this point a slight complication arises. In principle we want to assign a unique symbol to each pseudodifferential operator in U. But if $a(x,\xi) \in S^m(U)$ and if A is the operator defined by $a(x,\xi)$ via (I.13) (operators such as A make up a space which we denote by $OPS^m(U)$), any pseudodifferential operator which differs from A by a regularizing operator (cf. Def. I.1) must be assigned the same symbol $a(x,\xi)$. And conversely if we assume that A (given by (I.13)) is itself regularizing, then any regularizing operator will have $a(x,\xi)$ as symbol - as well as any other symbol which similarly defines a regularizing operator (via formulae of the kind (I.13)). This makes it clear that we must reason modulo symbols which define regularizing operators. These form a space which we denote by $S^{-\infty}(U)$ and which can be shown to be identical with the intersection of the spaces $S^m(U)$ as m ranges over $\mathbb{R}$ ([6], Th. 2.8). We shall write

(I.20) $$\dot{S}^m(U) = S^m(U)/S^{-\infty}(U) , \qquad \dot{S}(U) = \bigcup_m \dot{S}^m(U) .$$

<u>Definition I.3.- Equivalence classes belonging to</u> $\dot{S}^m(U)$ <u>will be called asymptotic symbols of degree</u> m <u>in</u> U.

In a sense the equivalence class modulo $S^{-\infty}(U)$ of a symbol $a(x,\xi)$ characterizes its "growth" as $|\xi| \to +\infty$. In particular, if $a(x,\xi) = 0$ for large $|\xi|$, its class mod $S^{-\infty}(U)$ is zero.

The most common method of constructing asymptotic symbols is based on use of <u>formal symbols</u>:

<u>Definition I.4.- By a formal symbol of degree</u> m <u>in</u> U <u>we mean a sequence</u>

<u>of symbols</u> $a_j(x,\xi)$ <u>in</u> U, <u>of respective degrees</u> m_j ($j = 0, 1, \ldots$), <u>such that</u> $m_0 = m$, <u>the</u> m_j <u>decrease and tend to</u> $-\infty$.

A formal symbol $\{a_j\}_{j=0,1,\ldots}$ is usually denoted as a series (actually, a "formal series"), $\sum_{j=0}^{+\infty} a_j$. It can easily be shown, then, that there exist cut-off functions $\chi_j(x,\xi) \in C^\infty(U \times \mathbb{R}^n)$ such that the modified series

$$\text{(I.21)} \qquad \sum_{j=0}^{+\infty} \chi_j(x,\xi) a_j(x,\xi)$$

actually converges in $S^m(U)$ (commonly one selects a sequence of compact subsets K_j of U, with K_j contained in the interior of K_{j+1} , exhausting U, and a sequence of numbers $R_j \nearrow +\infty$, and chooses the χ_j so as to have $\chi_j(x,\xi) = 0$ for $x \notin K_{j+1}$ or $|\xi| < R_j$, $\chi_j(x,\xi) = 1$ for x in K_j and $|\xi| > R_{j+1}$). Of course, the choice of the cut-offs χ_j can be varied in many ways; the important fact is that the corresponding "true" symbols (I.21) differ by elements of $S^{-\infty}(U)$. Thus we have the "transitions"

(I.22) formal symbols $\rightsquigarrow$ asymptotic symbols $\rightsquigarrow$ symbols

(the last arrow consisting of selecting arbitrarily a representative of any given asymptotic symbol). An important class of formal symbols is the one in the following

<u>Definition I.5.- By a classical symbol of degree m we mean a formal symbol</u> $\sum_j a_j(x,\xi)$ <u>such that, for each</u> $j = 0, 1, \ldots$, $a_j(x,\xi)$ <u>is a</u> C^∞ <u>function of</u> (x,ξ) <u>in</u> $U \times \mathbb{R}^n$, <u>positive-homogeneous of degree</u> $m - j$

with respect to ξ (i. e., $a_j(x,\rho\xi) = \rho^{m-j}\, a_j(x,\xi)$, $\rho > 0$) for large $|\xi|$.

It is customary to extend by homogeneity to all values of $\xi \neq 0$ every individual term $a_j(x,\xi)$ in the classical symbol $\sum_j a_j$, and therefore relinquish the demand that these terms be C^∞ in the whole of $U \times \mathbb{R}^n$; they now are C^∞ in $U \times (\mathbb{R}^n \setminus \{0\})$.

Definition I.6.- If $\sum_j a_j$ is a classical symbol in U, and if P is a pseudodifferential operator in U having it as its (formal) symbol, the term of degree m, $a_0(x,\xi)$, is called the principal symbol of P. We shall denote it by $\sigma(P)$.

The notion of principal symbol can be generalized to symbols other than the classical ones, but we shall not do so here.

So far we know that certain pseudodifferential operators have symbols and that, in this case (and provided we deal with asymptotic symbols), their symbols are uniquely defined (thanks to Th. 2.8 of [6]). We come now to the question of assigning a symbol to every pseudodifferential operator. This is easily done: If $P \in \Psi^m(U)$, P is congruent mod $\Psi^{-\infty}(U)$ to an element $A \in OPS^m(U,U)$; let then $a(x,y,\xi) \in S^m(U,U)$ be an amplitude defining A as in (I.12). Let us write, for any $N \in \mathbb{Z}_+$,

$$a(x,y,\xi) = \sum_{|\alpha|\leq N} \frac{1}{\alpha!}(x-y)^\alpha \partial_y^\alpha a(x,x,\xi) + \sum_{|\alpha|=N+1} (x-y)^\alpha a_\alpha(x,y,\xi) . \tag{I.23}$$

Integration by parts yields at once that

$$\int e^{i(x-y)\cdot\xi}(x-y)^\alpha b(x,y,\xi)\, d\xi = \int e^{i(x-y)\cdot\xi}(-D_\xi)^\alpha b(x,y,\xi)\, d\xi .$$

This in turn implies that

$$\text{(I.24)} \qquad Au(x) = (2\pi)^{-n} \int e^{ix\cdot\xi} \Big(\sum_{|\alpha| \leq N} \frac{1}{\alpha!} (-D_\xi)^\alpha \partial_y^\alpha a(x,x,\xi) \Big) \hat{u}(\xi)\, d\xi$$

$$+ (2\pi)^{-n} \iint e^{i\xi\cdot(x-y)}\, r_N(x,y,\xi)\, u(y)\, dy\, d\xi \ ,$$

where

$$\text{(I.25)} \qquad r_N(x,y,\xi) = \sum_{|\alpha| = N+1} (-D_\xi)^\alpha a_\alpha(x,y,\xi) \ \in \ S^{m-N-1}(U,U) \ .$$

From this we deduce easily that

$$\text{(I.26)} \qquad \exp(-D_\xi \partial_y) a(x,y,\xi) \Big|_{y=x} = \sum_{\alpha \in \mathbb{Z}_+^n} \frac{1}{\alpha!} (-D_\xi)^\alpha \partial_y^\alpha a(x,x,\xi)$$

is a formal symbol for P. This in turn defines an asymptotic symbol (Def I.3) - the symbol of P. We conclude this part of the present section by recalling how the algebraic operations on pseudodifferential operators reflect on their symbols (we shall solely deal with asymptotic symbols, while omitting the adjective asymptotic: when we call symbol a formal series of symbols, we mean the asymptotic symbol defined by the formal symbol in question).

Theorem I.6.- Let $a(x,\xi)$ denote the symbol of $P \in \dot{\Psi}^m(U)$. Then that of ${}^tP$ is $\sum_\alpha \frac{1}{\alpha!}(-1)^{|\alpha|}(\partial_x^\alpha D_\xi^\alpha a)(x,-\xi)$, and that of the adjoint P^* is $\sum_\alpha \frac{1}{\alpha!} \partial_x^\alpha D_\xi^\alpha \overline{a(x,\xi)}$.

Let $Q \in \dot{\Psi}^{m'}(U)$ with symbol $b(x,\xi)$. The symbol of PoQ is equal to

$$\text{(I.27)} \qquad (a \odot b)(x,\xi) = \sum_{\alpha \in \mathbb{Z}_+^n} \frac{1}{\alpha!} D_\xi^\alpha a(x,\xi)\, \partial_x^\alpha b(x,\xi) \ .$$

Corollary I.2.- Let $c(x,\xi)$ be the symbol of the commutator $[P,Q] =$ PoQ - QoP . Then $c(x,\xi) - \frac{1}{i}\{a,b\}(x,\xi) \in \dot{S}^{m+m'-2}(U)$.

We have denoted by $\{a,b\}$ the Poisson bracket of a and b :

$$\{a,b\} = \sum_{j=1}^{n} \frac{\partial a}{\partial \xi_j}\frac{\partial b}{\partial x_j} - \frac{\partial a}{\partial x_j}\frac{\partial b}{\partial \xi_j} . \tag{I.28}$$

<u>Corollary I.3</u>.- <u>If</u> $P \in \dot{\Psi}^m(U)$, $Q \in \dot{\Psi}^{m'}(U)$, <u>then</u> $PoQ \in \dot{\Psi}^{m+m'-1}(U)$.

Let us now return to the diagram (I.17) and denote by $J_\phi(x)$ the <u>Jacobian matrix</u> of the diffeomorphism ϕ at the point x. Let us denote by y the variable point in V, and by ${}^tJ_\phi^{-1}(y)$ the transpose of the inverse (<u>i. e.</u>, the <u>contragredient</u>) of $J(\phi^{-1}(y))$.

<u>Theorem I.7</u>.- <u>Let</u> $a(x,\xi)$ <u>denote the symbol of</u> $P \in \dot{\Psi}^m(U)$, $a^\phi(y,\eta)$ <u>that of</u> $P^\phi \in \dot{\Psi}^m(V)$. <u>We have</u>

$$a^\phi(y,\eta) \equiv a(\phi^{-1}(y), {}^tJ_\phi^{-1}(y)\eta) \mod \dot{S}^{m-1}(V). \tag{I.29}$$

The full significance of Th. I.7 is best understood when defining the symbols of a pseudodifferential operator in a manifold X, P. Let $(V,x_1,\ldots,x_n)$ be an arbitrary local chart in X (thus n = dim X), P_V the compose (I.18). By means of the local coordinates $(x_1,\ldots,x_n)$ we may transfer P_V from V to an open subset U of $\mathbb{R}^n$; the transferred pseudodifferential operator $P_U^\#$ has a symbol (belonging to $\dot{S}(U)$) which we define as <u>the symbol of</u> P <u>in the local chart</u> $(V,x_1,\ldots,x_n)$. It follows at once from Th. I.7 that the property that the symbol of P in any local chart is <u>classical</u> (Def. I.5) is independent of the choice of the local coordinates. We may therefore talk of classical pseudodifferential operators in a manifold. If P is a classical pseudodifferential operator of order m in X, in each local chart $(V,x_1,\ldots,x_n)$, it has a principal symbol (Def. I.6), $\sigma(P)(x,\xi)$. If we change coordinates in V and call $y_1,\ldots,y_n$, the new coordinates, and $\xi_1,\ldots,\xi_n$ the associated cocoordinates, we derive from (I.29) that

$$\sigma(P)(y,\eta) = \sigma(P)(x, {}^tJ_\phi^{-1}(x)\eta), \tag{I.30}$$

where $y = \phi(x)$. The meaning of (I.30) is that $\sigma(P)$ defines a bona fide function on the cotangent bundle T^*X over X or, more accurately, on the complement of the zero section, $T^*X \setminus 0$, in that bundle. Actually, because of the assumed positive-homogeneity (of degree m) of $\sigma(P)(x,\xi)$ with respect to ξ, we may view it as a C^∞ function on the cosphere bundle S^*X over X.

One of the great advantages of pseudodifferential operators is that they enable us to microlocalize. Let Γ be a conic open subset of $T^*X \setminus 0$: here conic refers to the fact that, if (x,ξ) belongs to Γ, so does $(x, \rho\xi)$ whatever $\rho > 0$. We shall call base of Γ its canonical image in the cosphere bundle S^*X. We shall denote by $\Psi_c^m(\Gamma)$ the space of pseudodifferential operators in X whose asymptotic symbol in any local chart has its support in an "open cone" whose base is relatively compact in Γ (this means that some true symbol representing the asymptotic symbol has this property). On the other hand we shall say that two operators P, Q $\in \Psi(X)$ are equivalent in Γ, and write $P \sim Q$ in Γ, if their asymptotic symbols in any local chart $(V, x_1, \ldots, x_n)$ are equal in the portion of Γ which lies above V. If $P \sim 0$ in Γ, we say that P is regularizing in Γ. It is clear that one may define the sheaf of germs of pseudodifferential operators over $T^*X \setminus 0$ by means of the presheaf $\{\tilde{\Psi}(\Gamma)\}$, where Γ ranges over all possible conic open subsets of $T^*X \setminus 0$, and $\tilde{\Psi}(\Gamma)$ denotes the space of equivalence classes modulo the relation $P \sim Q$ in Γ.

Definition I.7.- We shall say that a distribution u in X belongs to $C^\infty(\Gamma)$ (resp., to $H^s_{loc}(\Gamma)$) if $Pu \in C^\infty(X)$ (resp., $Pu \in H^s_{loc}(X)$) whatever the (properly supported) pseudodifferential operator P in X belonging to $\Psi_c^0(\Gamma)$.

Definition I.8.- Let $u \in \mathcal{D}'(X)$. We call wave-front set of u and denote by $WF(u)$ the complement of the union of all the conic open subsets Γ of $T^*X \setminus 0$ such that $u \in C^\infty(\Gamma)$.

Thus $WF(u)$ is a conic closed subset of u .

(I.31) The projection of $WF(u) \in T^*X \setminus 0$ into the base manifold, X, is exactly equal to sing supp u.

Theorem I.8.- Let $P \in \Psi^m(X)$ be properly supported. Given any distribution u in X such that $u \in H^s_{loc}(\Gamma)$ we have $Pu \in H^{s-m}_{loc}(\Gamma)$. If u belongs to $C^\infty(\Gamma)$ so does Pu .

Corollary I.4.- $\forall\, u \in \mathcal{D}'(X)$, $WF(Pu) \subset WF(u)$.

Let us denote by $\natural'(\Gamma)$ the quotient of $\mathcal{D}'(X)$ (viewed as a linear space without topology) modulo the subspace formed by the distributions in X which belong to $C^\infty(\Gamma)$. The sheaf defined by the presheaf $\{\natural'(\Gamma)\}$ (over $T^*X \setminus 0$ or, more correctly, over the cosphere bundle S^*X) is sometimes called the sheaf of microfunctions in X.

Lastly a few words about elliptic pseudodifferential operators.

First let U be an open subset of $\mathbb{R}^n$. A symbol $a(x,\xi) \in S^m(U)$ is said to be elliptic if there is a symbol $b(x,\xi) \in S^{-m}(U)$ such that $a(x,\xi)\cdot b(x,\xi) \equiv 1$ in $U \times \mathbb{R}^n$. An asymptotic symbol of degree m will be called elliptic if it has a representative which is an elliptic "true" symbol of degree m. These definitions transfer at once to any local chart $(V, x_1, \ldots, x_n)$ in the n-dimensional manifold X.

Definition I.8.- A pseudodifferential operator P of order m in X is said to be elliptic if its symbol in every local chart is elliptic.

If the symbol of P in every local chart is classical (i. e., if P is a classical pseudodifferential operator) we may make use of its principal symbol $\sigma(P)$ which is defined in the whole of $T^*X \setminus 0$ (and is positive-homogeneous of degree m). Then P <u>is elliptic if and only if</u> $\sigma(P)$ <u>does not vanish at any point of</u> $T^*X \setminus 0$.

<u>Theorem I.9</u>.- $P \in \Psi^m(X)$ <u>is elliptic if and only if there is</u> $Q \in \Psi^{-m}(X)$ <u>such that</u> $PQ = QP = I \mod \Psi^{-\infty}(X)$.

In order to construct the "approximate inverse" Q of Th. I.9 it suffices to construct its asymptotic symbol in each local chart of X. Thus it suffices to reason in the open set U of $\mathbb{R}^n$. The asymptotic symbol of Q is then constructed by first associating with Q a formal symbol, as follows: Let $\mathring{a}(x,\xi)$ denote the symbol of $P \in \Psi^m(U)$, $a(x,\xi) \in S^m(U)$ a representative of it such that $a(x,\xi)^{-1} \in S^{-m}(U)$. We then seek a formal symbol $b(x,\xi) = \sum_{j=0}^{+\infty} b_j(x,\xi)$ such as to have $a \odot b = 1$ (see (I.27)). This is achieved by taking:

$$(I.32)_0 \qquad b_0(x,\xi) = a(x,\xi)^{-1} ,$$

$$(I.32)_{j>0} \qquad b_j(x,\xi) = -\, a(x,\xi)^{-1} \sum_{1 \le |\alpha| \le j} \frac{1}{\alpha!} D_\xi^\alpha a(x,\xi)\, \partial_x^\alpha b_{j-|\alpha|}(x,\xi) .$$

From Cor. I.4 and Th. I.9 we derive at once:

<u>Corollary I.5</u>.- <u>If</u> $P \in \Psi^m(X)$ <u>is elliptic</u>,

$$(I.33) \qquad WF(Pu) = WF(u) , \qquad \forall\, u \in \mathcal{D}'(X).$$

Of course (see (I.31)) (I.33) has the implication that sing supp Pu = sing supp u for all distributions u in X. It should be noted that (I.33) is not the exclusive property of elliptic pseudodifferential operators; certain nonelliptic operators also enjoy it.

In practice we shall want to deal with functions and distributions whose values lie in Banach spaces (most often, in finite dimensional ones) and therefore deal with pseudodifferential operators valued in spaces of the Kind $L(E;F)$, the (Banach space) of bounded linear operators $E \to F$. Most statement that precede extend routinely, but some do not. We shall indicate some points that require care.

First of all, when the distributions are valued in a Banach space E which is not a Hilbert space, the definition of Sobolev spaces $H^s(\mathbb{R}^n;E)$ is the same as that in the scalar case, but does not make out of $H^s(\mathbb{R}^n;E)$ a Hilbert space - only a Banach space, and although $H^0(\mathbb{R}^n;E)$ is identical to $L^2(\mathbb{R}^n;E)$ as a topological vector space, this might not be so as a normed space, since Plancherel's formula is in general not valid (it is valid whenever E is a Hilbert space). Unless E itself is reflexive, $H^s(\mathbb{R}^n;E)$ will not be, nor will $H^s_{loc}(X;E)$ nor $H^s_c(X;E)$ be reflexive.

In dealing with pseudodifferential operators valued in $L(E)$, one must pay particular attention to the implication of noncommutativity - not only of the scalar pseudodifferential operators, but of the elements of $L(E)$. For instance, if $\dim E > 1$, Cor. I.3 will not be any more valid (in general): this is due to the fact that, if a and b are two symbols valued in $L(E)$, of degrees m and m' respectively, the commutator $a \circledcirc b - b \circledcirc a$ (see (I.27)) has degree $m + m'$ and not (in general) $m + m' - 1$, due to the fact that their usual commutator, $a(x,\xi)b(x,\xi) - b(x,\xi)a(x,\xi)$, does not necessarily vanish identically.

Finally we recall that a pseudodifferential operator with values in $L(E)$ is said to be elliptic when its principal symbol is <u>invertible</u> at every point of $T^*X \setminus 0$.

II. THE GENERALIZED HEAT EQUATION AND ITS PARAMETRIX

Throughout this chapter X will be, as before, a C^∞ manifold countable at infinity; $n = \dim X$; t will be the variable in the real line $\mathbb{R}$ (most often, in the closed half-line $\bar{\mathbb{R}}_+$). By m we denote some number > 0 , which, in the most significant applications, is equal either to <u>one</u> or to <u>two</u>; T will be some number > 0.

We shall deal with functions and distributions valued in a Hilbert space H (over $\mathbb{C}$). In the application H will be finite dimensional but there is no added complication by not requiring $\dim H < +\infty$. The norm in H will be denoted by $|\ \ |_H$, whereas the operator norm in L(H), the space of bounded linear operators in H, will be denoted by $\|\ \ \|$. The inner product in H will be denoted by $(\ ,\ )_H$.

Our basic ingredient is a pseudodifferential operator of order m in X, A(t), valued in L(H), depending smoothly on t in $[0,T[$. This means that in every local chart $(\Omega, x_1, \dots, x_n)$ A(t) is <u>congruent modulo regularizing operators which are</u> C^∞ <u>functions of</u> t (what we shall always shorten into "equivalent" and symbolize by $\sim$; here the regularizing operators must be valued in L(H)) to an operator

$$\text{(II.1)} \qquad A_\Omega(t)u(x) = (2\pi)^{-n} \int e^{ix\cdot\xi} a_\Omega(x,t,\xi)\hat{u}(\xi)d\xi \ , \quad u \in C_c^\infty(\Omega;H),$$

where

(II.2) $a_\Omega(x,t,\xi)$ <u>is a</u> C^∞ <u>function of</u> $t \in [0,T[$ <u>valued in</u> $S^m(\Omega;L(H))$.

In the forthcoming we shall often drop the subscript Ω and refer to $a(x,t,\xi)$ as the symbol of A(t) in the chart $(\Omega, x_1,\dots,x_n)$ although this name should be reserved for the class of $a(x,t,\xi)$ mod $S^{-\infty}(\Omega;L(H))$.

We are interested in solving the following initial value problem:

(II.3) $\frac{dU}{dt} - A(t)\circ U \sim 0$ *in* $X \times [0,T[$,

(II.4) $U|_{t=0} \sim I$, *the identity of* H, *in* X.

In principle the solution $U(t)$ should be an equivalence class, modulo regularizing operators in X depending smoothly on t, of continuous linear operators $\mathcal{E}'(X;H) \to \mathcal{D}'(X;H)$ depending smoothly on t (in $[0,T[$). But without additional hypotheses about $A(t)$ there is no reason why such a solution should exist. We are going to make a hypothesis on $A(t)$ which not only will insure that U does exist, but also that it possesses a convenient integral representation. Observe that when $X = \mathbb{R}^n$ and $A(t) = \Delta_x$, the Laplace operator in n variables, in which case the symbol of $A(t)$ is $-|\xi|^2$, Eqq. (II.3)-(II.4) define the parametrix in the forward Cauchy problem for the heat equation $\frac{\partial U}{\partial t} - \Delta_x U = 0$. In general we shall make the following hypothesis:

(II.5) *Let* $(\Omega, x_1,\dots,x_n)$ *be any local chart in* X. *There is a symbol* $a(x,t,\xi)$ *satisfying* (II.2), *and defining via* (II.1) *the operator* $A_\Omega(t)$ *congruent to* $A(t)$ *modulo regularizing operators in* Ω *depending smoothly on* $t \in [0,T[$, *such moreover that*

(II.6) *to every compact subset* K *of* $\Omega \times [0,T[$ *there is a compact subset* K' *of the open half-plane* $\mathbb{C}_- = \{z \in \mathbb{C} \ ; \ \mathrm{Re}\, z < 0\}$ *such that*

(II.7) $$zI - a(x,t,\xi)/(1 + |\xi|^2)^{m/2} \ : H \to H$$

is a bijection (hence also a homeomorphism), *whatever* (x,t) *in* K, ξ *in* $\mathbb{R}^n$, z *in* $\mathbb{C} \setminus K'$.

We may now state the main result of this chapter:

Theorem II.1.- Under Hypothesis (II.5) the problem (II.3)-(II.4) has a solution U(t) which is a function of $t \in [0,T[$ valued in $\dot{\Psi}^0(X;L(H))$ (cf. (I.19)). There is a representative of the equivalence class U(t) with the following property:

In each local chart $(\Omega, x_1,\ldots,x_n)$ of X the representative in question is equivalent to an element $U_\Omega(t)$ of $OPS^0(\Omega;L(H))$ given by

(II.8) $$U_\Omega(t)u(x) = (2\pi)^{-n}\int e^{ix\cdot\xi}\,\mathcal{U}_\Omega(x,t,\xi)\hat{u}(\xi)d\xi\,,\quad u\in C_c^\infty(\Omega;H),$$

whose symbol $\mathcal{U}_\Omega$ has the following properties:

(II.9) $\mathcal{U}_\Omega$ is a C^∞ map $\Omega \times [0,T[\times \mathbb{R}^n \to L(H)$;

(II.10) to every compact subset $\mathcal{K}$ of $\Omega \times [0,T[$, to every pair of n-tuples $\alpha, \beta \in \mathbb{Z}_+^n$ and to every pair of integers $r, N \geq 0$ there is a constant $C > 0$ such that, whatever (x,t) in $\mathcal{K}$, $\xi \in \mathbb{R}^n$,

(II.11) $$\|\partial_x^\alpha \partial_\xi^\beta \partial_t^r\, \mathcal{U}_\Omega(x,t,\xi)\| \leq C\, t^{-N}\, (1 + |\xi|)^{rm-|\beta|-Nm}\,.$$

Any C^∞ function of t in $[0,T[$ valued in the space of continuous linear mappings $\mathcal{E}'(X;H) \to \mathcal{D}'(X;H)$ which satisfies (II.3)-(II.4) belongs to the equivalence class U(t).

It follows from (II.11) that, for $t > 0$, $\mathcal{U}_\Omega(x,t,\xi)$ belongs to $S^{-\infty}(\Omega;L(H))$, i. e. the operator (II.8) is regularizing, in other words, the equivalence class U(t) is zero. This generalizes the well-known property of the parametrix of the heat equation.

Proof of Th. II.1: A. Existence of the parametrix U(t)

It suffices to reason in the (generic) local chart $(\Omega, x_1,\ldots,x_n)$ and patch the $U_\Omega(t)$ together afterwards, by means of a smooth partition of unity in X. Thus we construct the symbol $\mathcal{U}_\Omega$; actually we construct a formal symbol (Def. I.4)

$$\mathcal{U}(x,t,\xi) = \sum_{j=0}^{+\infty} \mathcal{U}_j(x,t,\xi) , \tag{II.12}$$

out of which a true symbol can later on be constructed, by using cut-offs as indicated in Ch. I. We take the operator (II.1) to be the operator $A_\Omega(t)$ in (II.5) and omit the subscripts Ω ; we do not distinguish any more between $A(t)$ and $A_\Omega(t)$, which we also denote by $a(x,t,D_x)$ ($a(x,t,\xi)$ is it symbol). Reasoning formally we write

$$\left[\frac{d}{dt} - A(t)\right]U(t)u = \tag{II.13}$$

$$(2\pi)^{-n}\int e^{ix\cdot\xi}\left[\frac{\partial}{\partial t} - a(x,t,D_x+\xi)\right]\mathcal{U}(x,t,\xi)\hat{u}(\xi)d\xi ,$$

and require, for $0 \leq t < T$,

$$\frac{\partial\mathcal{U}}{\partial t} - a(x,t,D_x+\xi)\mathcal{U} = \frac{\partial\mathcal{U}}{\partial t} - \sum_{\alpha\in\mathbb{Z}_+^n} \frac{1}{\alpha!}\partial_\xi^\alpha a(x,t,\xi)D_x^\alpha\mathcal{U} = 0 , \tag{II.14}$$

which may be rewritten, with the notation (I.27),

$$\frac{\partial\mathcal{U}}{\partial t} - a(x,t,\xi)\circledcirc\mathcal{U} = 0 , \quad 0 \leq t < T . \tag{II.15}$$

This eq. (II.15) is the "translation" of (II.3); as for (II.4) it translates into

$$\mathcal{U}(x,0,\xi) = I \text{ (the identity of H).} \tag{II.16}$$

By availing ourselves of the basic hypothesis, (II.5), we are going to obtain $\mathcal{U}(x,t,\xi)$ in the form

$$\mathcal{U}(x,t,\xi) = (2\pi i)^{-1}\oint_\gamma e^{\rho tz} k(x,t,\xi;z)\, dz , \tag{II.17}$$

where k is a suitable formal symbol of degree zero, valued in L(H), depending holomorphically on the complex variable z in an open neighborhood of the integration contour γ provided (x,t) remains in a given compact subset of $\Omega\times[0,T[$. We have used the notation $\rho = \rho(\xi) = (1+|\xi|^2)^{\frac{1}{2m}}$, and shall continue to use it henceforth.

We select arbitrarily a relatively compact open subset Ω_o of Ω, a number T_o, $0 < T_o < T$, and take the compact set K in (II.6) to be the closure of $\mathcal{O} = \Omega_o \times]0,T_o[$. We take the compact subset K' of $\mathbb{C}_-$ in (II.6) accordingly, and denote by M the maximum norm of the inverse of the mapping (II.7) as (x,t) ranges over K, ξ over $\mathbb{R}^n$ and z over a simple closed smooth curve γ winding around K' in $\mathbb{C}_- \setminus K'$.

Since $a \odot (e^{\rho tz}k) = e^{\rho tz}(a \odot k)$ we may rewrite Eq. (II.15) as

(II.18) $$\oint_\gamma e^{\rho tz} \mathcal{L} k(z,t,\xi;z)\, dz = 0, \quad \text{where:}$$

(II.19) $$\mathcal{L} k = \frac{\partial k}{\partial t} + \rho z k - a(x,t,\xi) \odot k .$$

We are going to solve (in the sense of formal symbols) the equation

(II.20) $$\mathcal{L} k = \rho I,$$

which implies at once (II.18). It will turn out that the (unique) formal symbol k satisfying (II.20) will also satisfy

(II.21) $$(2\pi i)^{-1} \oint_\gamma k(x,t,\xi;z)dz = I, \quad \forall (x,t) \in \mathcal{O},\ \xi \in \mathbb{R}^n,$$

which, for t = 0, is nothing else but (II.16).

Solution of (II.20): We rewrite (II.20) as follows:

(II.22) $$k = E\left[I - \rho^{-1}\left(\frac{\partial k}{\partial t} - a \odot k + ak\right)\right],$$

setting $E = \left[zI - \rho^{-1}a(x,t,\xi)\right]^{-1}$ (inverse in L(H)). We note that

(II.23) $$\rho^{-1}(a \odot k - ak) = \sum_{\alpha \neq 0} \frac{1}{\alpha!} \rho^{-1} \partial_\xi^\alpha a\, D_x^\alpha k$$

has degree $\leq$ deg k - 1. We solve (II.22) by taking $k = \sum_{j=0}^{+\infty} k_j$ and requiring:

$(II.24)_0$ $$k_0 = E,$$

$(II.24)_{j>0}$ $$k_j = -E\,\rho^{-1}\left[\frac{\partial}{\partial t}k_{j-1} - \sum_{1\leq|\alpha|\leq j} \frac{1}{\alpha!}\partial_\xi^\alpha a\, D_x^\alpha k_{j-|\alpha|}\right].$$

By induction on j we see easily that

(II.25) $m_j = \deg k_j \leqslant -j \inf(1,m)$,

which implies $\sum_j k_j$ indeed defines a formal symbol (since $m > 0$).

Furthermore,

(II.26) If $j \geqslant 1$, k_j is a finite sum of terms of the form $Eb_1E\ldots b_rE$ with r varying from term to term but always remaining $\geqslant 2$, and with each b_i a C^∞ function of t in $[0,T[$ valued in $S^{d_i}(\Omega;L(H))$ independent of z ($i = 1,\ldots, r$), such moreover that $d_1 + \ldots + d_r \leqslant m_j$.

According to (II.26), therefore,

(II.27) To every relatively compact open subset Ω_0 of Ω and to every number T_0, $0 < T_0 < T$, there is a compact subset K' of $\mathbb{C}_-$ such that, for each $j = 0, 1,\ldots$, $k_j(x,t,\xi;z)$ is a C^∞ function of (t,z) in $[0,T_0[\times (\mathbb{C} \setminus K')$, holomorphic with respect to z, valued in $S^{m_j}(\Omega_0;L(H))$.

Proof of (II.21): Fix arbitrarily (x,t) in $\mathcal{O}$ and ξ in $\mathbb{R}^n$; then $a_0 = \rho^{-1}a(x,t,\xi)$ is a bounded linear operator $H \to H$ and so are the b_i in (II.26). Writing $E(z) = (zI - a_0)^{-1}$ and keeping in mind that γ winds around the spectrum of a_0 we get once:

$$(2\pi i)^{-1} \oint_\gamma E(z)dz = I \,, \qquad \oint_\gamma E(z)b_1E(z)\ldots b_rE(z)\, dz = 0 \ (\text{if } r \geqslant 1),$$

whence (II.21) (by using (II.26)).

Estimates of the symbols $\mathcal{U}_j(x,t,\xi) = (2\pi i)^{-1} \oint_\gamma e^{\rho t z} k_j(x,t,\xi;z)dz$

By (II.27) we see that this formula defines $\mathcal{U}_j(x,t,\xi)$ for all (x,t) in $\Omega \times [0,T[$: indeed, it does for (x,t) in $\mathcal{O}$, but if we replace $\mathcal{O}$ by a larger open set $\mathcal{O}_1$ we might replace γ by a different contour. If we then restrict back (x,t) to $\mathcal{O}$ it follows from the Cauchy integral theorem that we recover the same value as before. We note that, if $z \in \gamma$,

$$\text{(II.28)} \qquad \left|\partial_\xi^\beta \partial_t^r (e^{\rho t z})\right| \leq \text{const.}(1+|\xi|)^{-|\beta|} \rho^r \sum_{\ell=0}^{|\beta|} (t\rho)^\ell e^{\rho t \operatorname{Re} z}$$

$$\leq \text{const.}(1+|\xi|)^{-|\beta|} \rho^{r-N} t^{-N} \sum_{\ell=N}^{|\beta|+N} (\rho t)^\ell e^{-c_o \rho t}$$

$$\leq \text{const. } t^{-N} (1+|\xi|)^{-|\beta|} \rho^{r-N}$$

(we have availed ourselves of the fact that $\operatorname{Re} z \leq - c_o < 0$ on γ).

On the other hand we derive from (II.27); for (x,t) in $\mathcal{O}$ and z in γ,

$$\text{(II.29)} \qquad \left\| \partial_x^\alpha \partial_\xi^\beta \partial_t^r k_j(x,t,\xi\,;z)\right\| \leq \text{const. } (1+|\xi|)^{m_j -|\beta|}.$$

By combining (II.28) and (II.29), and applying Leibniz formula, we get

$$\text{(II.30)} \qquad \left\| \partial_x^\alpha \partial_\xi^\beta \partial_t^r \mathcal{U}_j(x,t,\xi)\right\| \leq C\, t^{-N}(1+|\xi|)^{m_j+(r-N)m-|\beta|}$$

for all (x,t) in $\mathcal{O}$, ξ in $\mathbb{R}^n$. This implies (II.11).

B. Uniqueness of the parametrix

The uniqueness of the parametrix, needless to say in the sense of equivalence class modulo regularizing operators, follows from various standard considerations which are of interest in their own right, and which we now go into rapidly.

First of all we did not have to solve Equation (II.3) while prescribing the value of the solution at time $t = 0$. We could have solved

$$\text{(II.31)} \qquad \frac{dU}{dt} - A(t)\circ U \sim 0 \quad \underline{\text{in}}\ X \times [t', T[\,, \qquad U\big|_{t=t'} = I \ \underline{\text{in}}\ X,$$

where t' is any number such that $0 \leq t' < T$. By the same procedure as in Part A we can find a solution $U(t,t')$ having a representative which, in any local chart $(\Omega, x_1,\dots,x_n)$, is equivalent to an operator $U_\Omega(t,t')$ defined by

$$\text{(II.32)} \qquad U_\Omega(t,t')u(x) = (2\pi)^{-n}\int e^{ix\cdot\xi}\, \mathcal{U}_\Omega(x,t,t',\xi)\hat{u}(\xi)d\xi,$$

with

(II.33) $$\mathcal{U}_\Omega(x,t,t',\xi) = (2\pi i)^{-1} \oint_\gamma e^{-(t-t')\rho z} k_\Omega(z,t,\xi;z)\, dz,$$

where k_Ω is the same symbol as in (II.17): in particular it is independent of t' . This is due to the validity of (II.21) where we may take $t = t'$. The contour of integration γ may also be taken to be the same as in Part A.

The solution of (II.31) enables us to solve the inhomogeneous Cauchy problem:

(II.34) $$\frac{\partial u}{\partial t} - A(t)u = f \quad \text{in } X \times [0,T[\ , \quad u|_{t=0} = u_0 \quad \text{in } X .$$

Here f is an H-valued function or distribution in $X \times [0,T[$, u_0 an element of $\mathcal{D}'(X;H)$ (in all rigor we must reason modulo $C^\infty(X)$). If f is sufficiently regular with respect to t, say continuous, we may write

(II.35) $$u(t) = U(t)u_0 + \int_0^t U(t,t')f(t')dt' .$$

Next we look at the backward Cauchy problem for the adjoint equation. For each t in $[0,T]$ we denote by $A^*(t)$ the adjoint of the operator A(t) as an L(H)-valued pseudodifferential operator in X (in order to define the adjoint of A(t) we make use of a (strictly positive) density ϖ in X. Then, in any local chart $(\Omega, x_1,\dots,x_n)$, in which the symbol of A(t) is $a(x,t,\xi)$, a formal symbol or $A^*(t)$ is given by

(II.36) $$\sum_{\alpha\in\mathbb{Z}_+^n} \frac{1}{\alpha!} D_x^\alpha \partial_\xi^\alpha \left[\varpi(x)\, a(x,t,\xi)^*\right],$$

where $a(x,t,\xi)^*$ stands for the adjoint of $a(x,t,\xi)$ as a bounded linear operator on H. If we assume that $a(x,t,\xi)$ satisfies the basic hypothesis (II.6) so will any reasonable true symbol constructed out of (II.36). By duplicating the construction in Part A we can now construct an operator V(t,t') solution to

(II.37) $$\frac{dV}{dt} + A^*(t)\circ V \sim 0 \quad \text{in } X \times]0,t'] \ , \quad V|_{t=t'} \sim I \quad \text{in } X,$$

where t' is any number such that $0 < t' < T$. As a matter of fact, our hypothesis that A(t) is smooth up to t = 0 implies that we can construct a representative of V(t,t') which is C^∞ in the <u>closed</u> interval $[0,t']$. This is important in the applications.

Let then $V^*(t,t')$ denote the adjoint of V(t,t') or, rather, a properly supported representative of this equivalence class. By transposing (II.37) we get

$$\frac{dV^*}{dt'}(t',t) + V^*(t',t)\circ A(t') = R(t,t') \quad \underline{\text{in}}\ X \times [0,t], \tag{II.38}$$

$$V^*(t\ ,t) - I = R_0(t) \quad \underline{\text{in}}\ X, \tag{II.39}$$

with R, R_0 regularizing (and depending smoothly on t, t'. If we therefore define the operator G by the formula

$$Gf(t) = \int_0^t V^*(t',t)\ f(t')dt' \ , \tag{II.40}$$

and suppose that (II.34) holds we obtain, by integration by parts,

$$Gf(t) = V^*(t,t)u(t) - V^*(0,t)u_0 - \tag{II.41}$$
$$- \int_0^t \left[\frac{dV^*}{dt'}(t',t) + V^*(t',t)\circ A(t')\right]u(t')dt' \ .$$

(At this point it is advisable to use a properly supported representative of A(t) so that both R and R_0 in (II.38), (II.39) are properly supported). Eq. (II.4.1) can be rewritten as

$$u(t) = G\left[\frac{\partial u}{\partial t} - A(t)u\right] + V^*(0,t)u(0) - R_0(t)u(t) + \tag{II.42}$$
$$+ \int_0^t R(t,t')u(t')\ dt' \ .$$

We may now easily prove the uniqueness of the parametrix. Let U(t) be a representative of the parametrix constructed in Part A, $U_1(t)$ one of another solution of (II.3)-(II.4); we assume that $U_1(t)$ is a C^∞ function of t in $[0,T[$ valued in the space of continuous linear mappings

$\mathcal{E}'(X;H) \to \mathcal{D}'(X;H)$. It suffices to put $u(t) = [U(t) - U_1(t)]w_0$ in (II.42), with $w_0 \in \mathcal{E}'(X;H)$ arbitrary, to conclude that $U(t) - U_1(t)$ is a C^∞ function of t in $[0,T[$ valued in $\Psi^{-\infty}(X;L(H))$.

The proof of Th. II.1 is complete.

Remark II.1.- The uniqueness of the parametrix $U(t)$ and the similar property for $U(t,t')$ have the consequence that

$$\text{(II.43)} \qquad U(t,t'') \sim U(t,t')U(t',t'') \quad \underline{\text{if}}\ 0 \le t'' \le t' \le t < T;$$

$$\text{(II.44)} \qquad U(t,t') \sim V^*(t',t) \quad \underline{\text{if}}\ 0 \le t' \le t.$$

Thus Formula (II.42) can be rewritten in the form

$$\text{(II.45)} \qquad u(t) = U(t)u(0) + \int_0^t U(t,t')\left[\frac{\partial u}{\partial t}(t') - A(t')u(t')\right]dt' - R_0(t)u(t) + \int_0^t R(t,t')u(t')dt' ,$$

for every $u \in C^\infty([0,T[;\mathcal{D}'(X;H))$. In turn the expression (II.45) implies the hypo-ellipticity of the Cauchy problem (II.34):

Theorem II.2.- Let Y be an open subset of X, u a C^∞ function of t in $[0,T[$ valued in $\mathcal{D}'(X;H)$.

Suppose that $u(0) \in C^\infty(Y;H)$ and that $\frac{\partial u}{\partial t} - A(t)u \in C^\infty(Y \times [0,T[;H)$. Then $u \in C^\infty(Y \times [0,T[;H)$.

Since pseudodifferential operators decrease wave-front sets (Cor. I.4) we could have replaced the open set $Y \subset X$, in Th. II.2, by any conic open subset of $T^*X \setminus 0$.

We shall now rapidly describe some properties of the parametrix $U(t)$ with very few hints about their proofs.

Link with the Laplace transform

Let $k(x,t,\xi;z)$ be the symbol in (II.17) and set

$$\text{(II.46)} \qquad k^{\#}(x,\xi;z) = \sum_{j=0}^{+\infty} \frac{1}{j!}(-\rho)^{-j} \frac{\partial^{2j} k}{\partial t^j \partial z^j}(x,0,\xi;z) \ ,$$

$$\text{(II.47)} \qquad \mathcal{U}^{\#}(x,t,\xi) = (2\pi i)^{-1} \oint_{\gamma} e^{\rho tz} k^{\#}(x,\xi;z)dz \ ,$$

where γ is the same contour as in (II.17). Note that $k^{\#}$ and $\mathcal{U}^{\#}$ are bona fide formal symbols. It is easy to check that $k^{\#}$ is the Laplace transform of $^{\#}$, in the following sense:

$$\text{(II.48)} \qquad k^{\#}(x,\xi;z) = \rho \int_0^{+\infty} e^{-\rho zt}\, \mathcal{U}^{\#}(x,t,\xi)\, dt \ .$$

For us the important thing is that

(II.49) $\mathcal{U}(x,t,\xi) - \mathcal{U}^{\#}(x,t,\xi)$ <u>is a</u> C^{∞} <u>function of</u> t <u>in</u> $[0,T[$ <u>with values in</u> $\Psi^{-\infty}(\Omega;L(H))$.

The motivation for this is that (reasoning formally)

$$\begin{aligned} \mathcal{U}(x,t,\xi) &= (2\pi i)^{-1} \sum_{j=0}^{+\infty} \oint_{\gamma} e^{\rho tz} \frac{t^j}{j!} \left(\frac{\partial}{\partial t}\right)^j k(x,0,\xi;z)\, dz \\ &= (2\pi i)^{-1} \sum_{j=0}^{+\infty} \frac{1}{j!} \oint_{\gamma} \left(\frac{1}{\rho}\frac{\partial}{\partial z}\right)^j (e^{\rho tz}) \quad \frac{\partial^j k}{\partial t^j}(x,0,\xi;z)\, dz \\ &= k^{\#}(x,\xi;z) \quad \text{by integrating by parts with respect to z.} \end{aligned}$$

One could have gone directly to $k^{\#}$ by performing a Laplace transform with respect to t on the operator $\frac{\partial}{\partial t} - \sum_j \frac{t^j}{j!} A^{(j)}(0)$, and obtained $\mathcal{U}^{\#}$ by inverse Laplace transform. By means of $\mathcal{U}^{\#}$ we obtain a representative $U^{\#}_{\Omega}(t)$ in the local chart $(\Omega, x_1,\dots,x_n)$ of the parametrix U(t) which happens to be a smooth function of t for <u>all</u> positive values of t, not just those $< T$ - this is not so surprising since we know we could extend U(t) to such values - just by setting U(t) = 0 (as an equivalence class) for all $t > 0$. The symbol (II.47) is now going to be used, in the study of the operator U^*U.

The operator U^*U

We assume that X is equipped with a density $\varpi > 0$ and use a properly supported representative of U(t), which we may and shall assume defined for all $t > 0$, in accordance with the last remark. We then regard $u_0 \mapsto U(t)u_0$ as a continuous linear mapping $\mathcal{D}'(X;H) \to C^\infty(\bar{R}_+;\mathcal{D}'(X;H))$, which we call U. It has an adoint, U^*, which we may regard as a continuous linear map $C_c^\infty(\bar{R}_+;\mathcal{D}'(X;H)) \to \mathcal{D}'(X;H)$, and which is given by

$$(U^*v)(x) = \int_0^{+\infty} U(t)^* v(x,t)\, dt, \qquad v \in C_c^\infty(X \times R_+;H), \tag{II.50}$$

where (for each $t \geq 0$) $U(t)^*$ is the adjoint of the pseudodifferential operator in X, U(t) (it is also properly supported). Let then $\zeta \in C_c^\infty(\bar{R}_+)$, $\zeta(t) = 1$ for $t < R$ $(R > 0)$ and let us set

$$K_\zeta = \int_0^{+\infty} \zeta(t) U(t)^* U(t) dt . \tag{II.51}$$

From the fact that U(t) is regularizing in X whatever $t > 0$ we derive at once that if $\zeta_1 \in C_c^\infty(\bar{R}_+)$, $\zeta_1(t) = 1$ for $t \in R_1$ ($R_1 > 0$ also) the difference $K_\zeta - K_{\zeta_1}$ is regularizing in X. We denote by U^*U the equivalence class of K_ζ modulo regularizing operators in X.

<u>Theorem II.3</u>.- <u>The compose</u> U^*U <u>is an elliptic pseudodifferential operator of order</u> $-m$ <u>in</u> X, <u>valued in</u> L(H), <u>formally self-adjoint</u>.

Actually, in proving Th. II.3, one can obtain detailed information about the symbol of U^*U in an arbitrary local chart $(\Omega, x_1, \ldots, x_n)$ of X. It is convenient to use the symbol (II.47) (we shall however omit the superscripts #). A formal symbol for $U(t)^*U(t) \in \Psi^0(X;L(H))$ in the local chart under consideration is obtained by taking (see Th. I.6)

$$\text{(II.52)} \qquad \sum_{\alpha,\beta\in\mathbb{Z}_+^n} \frac{1}{\alpha!\beta!} \partial_\xi^{\alpha+\beta} D_x^\alpha \left[\varpi(x)\mathcal{U}(x,t,\xi)^*\right] D_x^\beta \mathcal{U}(x,t,\xi) ,$$

and a formal symbol for U^*U is then obtained by integrating (II.52) from 0 to $+\infty$. This shows easily that U^*U is formally self-adjoint (i. e., its asymptotic symbol is self-adjoint, in an obvious sense). By taking into account (II.46) and (II.47) we see the symbol of U^*U we have just described differs from the following symbol

$$\text{(II.53)} \qquad \mathcal{K}(x,\xi) = \frac{\varpi(x)}{2\pi} \int_{-\infty}^{+\infty} \left(\left[a(x,0,\xi)-i\tau I\right]^{-1}\right)^* \left[a(x,0,\xi)-i\tau I\right]^{-1} d\tau ,$$

by a symbol of order $<-m$. Thus (II.53) plays a role analogous to that of principal symbol (at least in the local chart $(\Omega, x_1,\dots,x_n)$). Notice that

(II.54) $\quad \mathcal{K}(x,\xi)$ is elliptic (of order $-m$) positive-definite,

more precisely, if h is an arbitrary element of H, $x\in\Omega$, $\xi\in\mathbb{R}^n$,

$$\text{(II.55)} \qquad (\mathcal{K}(x,\xi)h,h)_H \geq \varpi(x)\,|h|_H^2/\left[\pi\|a(x,0,\xi)\|\right] .$$

Furthermore, if $a(x,0,\xi)$ and $a(x,0,\xi)^*$ commute in $L(H)$,

$$\text{(II.56)} \qquad \mathcal{K}(x,\xi) = -\,\varpi(x)\left[a(x,0,\xi)+a(x,0,\xi)^*\right]^{-1} .$$

It should be underlined that if $a(x,0,\xi)$ and $a(x,0,\xi)^*$ do not commute, the equality (II.56) might not hold and as a matter of fact, the right-hand side might not even exist, as the example $H=\mathbb{C}^2$, $a(x,t,\xi) = \begin{pmatrix} -1/2 & -1 \\ 0 & -1/2 \end{pmatrix}|\xi|$ shows. Instead of dealing with $U(t) = U(t,0)$ we could have dealt with $U(t,t')$, $0\leq t'\leq t<+\infty$. We could have thus proved that the class of pseudodifferential operators in X, which can be denoted by $\int_t^{+\infty} U(t,t')^* U(t,t')dt$, is elliptic of order $-m$. These results will now be exploited.

The local estimates

Let us denote by $U(t)$, $U(t,t')$ properly supported representatives of the equivalence classes so denoted above; we may also assume that $U(t)$ is defined for all $t \geqslant 0$, and $U(t,t')$ for all $t \geqslant t'$ $(t' \geqslant 0)$. By means of a positive density in X - or else by restricting the argument to a local chart $(\Omega, x_1, \ldots, x_n)$ - we may use Sobolev norms on X. For each s in $\mathbb{R}$ we denote the corresponding norm $\| \ \|_s$, and the inner product by $(\ , \ )_s$. In what follows $\mathcal{O}$ will denote a relatively compact open subset of X.

First of all, if u_0 is an arbitrary element of $C_c^\infty(\mathcal{O};H)$, we have

$$\text{(II.57)} \qquad \int_0^T \|U(t)u_0\|_0^2 \, dt = (K_T u_0, u_0)_0 , \quad K_T = \int_0^T U(t)^* U(t) dt .$$

But of course K_T is a representative of the class $U^* U$, and thus by Th. II.3 we derive that

$$\text{(II.58)} \qquad \int_0^T \|U(t)u_0\|_0^2 \, dt \leq \text{const.} \|u_0\|_{-m/2}^2 , \qquad u_0 \in C_c^\infty(X;H).$$

Next, let f be an arbitrary element of $C_c^\infty(\mathcal{O} \times \bar{R}_+;H)$

$$\text{(II.59)} \qquad \int_0^T \left\| \int_0^t U(t,t')f(t')dt' \right\|_0^2 dt \leq \text{const.} \int_0^T \|f(t)\|_{-m}^2 \, dt .$$

This follows from the extension of Th. II.3 to the equivalence class of operators which we have denoted by $\int_{t'}^{+\infty} U(t,t')^* U(t,t') dt$. Note that the constants, in (II.58) and (II.59), depend on the choice of $T > 0$ (but of course not on those of u_0 and f). By availing ourselves of (II.58) & (II.59) in conjunction with Formula (II.45) we obtain:

Theorem II.4.- To every relatively compact open subset $\mathcal{O}$ of X and to every number T' such that $0 < T' < T$, there is a constant $C > 0$ such that, for all u in $C_c^\infty(\mathcal{O} \times \bar{R}_+;H)$,

$$(\text{II.60})\qquad \int_0^{T'} \|u(t)\|_0^2\,dt \le C\Big\{\|u(0)\|^2_{-m/2} +$$

$$\int_0^{T'} \Big\|\frac{\partial u}{\partial t}(t) - A(t)u(t)\Big\|^2_{-m}\,dt + \int_0^{T'} \|u(t)\|^2_{s'}\,dt\Big\} \quad (s'<0).$$

All other estimates one might wish follow from this basic one, (II.60) : For instance, instead of estimating the zero-norm of $u(t)$, one can estimate its s-th norm; one can also estimate the s-th norm of the t-derivatives of u(t) (now, below, s' is any number $< s$; C depends on its choice):

$$(\text{II.61})\qquad \int_0^{T'} \|\partial_t^{k+1}u(t)\|_s^2\,dt \le C\Big\{\|u(0)\|^2_{s+km+m/2} +$$

$$+ \sum_{j\le k}\int_0^{T'} \Big\|\partial_t^j\Big[\frac{\partial u}{\partial t}(t) - A(t)u(t)\Big]\Big\|^2_{s+(k-j)m}\,dt + \int_0^{T'} \|u(t)\|^2_{s'}\,dt\Big\}.$$

Suitable estimates are also valid for the t-derivatives of negative order (i. e., integrals from 0 to t, iterated a number of times).

"Orthogonal projections on the kernel and cokernel"

Here we let $A(t)$, $U(t)$, $U(t,t')$ act on the space of "microfunctions"

$$(\text{II.62})\qquad C^\infty([0,T[;\dot{\mathcal{D}}'(X;H)) = C^\infty([0,T[;\mathcal{D}'(X;H))/C^\infty(X\times[0,T[;H),$$

and we let U^*U act on $\dot{\mathcal{D}}'(X;H) = \mathcal{D}'(X;H)/C^\infty(X;H)$ (by using properly supported representatives of all these classes of operators). Let us then denote by $(U^*U)^{-1}$ the inverse of U^*U ; by Th. II.3 it is an (elliptic) pseudodifferential operator of order m. We then form

$$(\text{II.63})\qquad P_0 = U(U^*U)^{-1}U^* \quad \underline{\text{acting on}}\ C^\infty([0,T[;\dot{\mathcal{D}}'(X;H)).$$

Theorem II.5.- In the sense of operators on the space (II.62) we have:

$$(\text{II.64})\qquad P_0 = P_0^* = P_0^2\ ;\quad P_0 = I \text{ on } \mathrm{Ker}\Big[\frac{\partial}{\partial t} - A(t)\Big]\ ;\ \Big[\frac{\partial}{\partial t} - A(t)\Big]\circ P_0 = 0\ .$$

One may paraphrase the equations (II.64) by saying that P_0 is an approximate orthogonal projection on the kernel of $\frac{\partial}{\partial t} - A(t)$ in the space of microfunctions, (II.62).

Lastly, for any $\dot{g}$ in (II.62), let us call $E_0\dot{g}$ the class in (II.62) of the distribution $\int_0^t U(t,t')g(t')dt'$, for some g in the class $\dot{g}$.

Theorem II.6.- With the preceding notation set $E = (I - P_0)E_0$. Then, in the sense of linear operators acting on the space (II.62), we have:

$$\text{(II.65)} \qquad \left[\frac{\partial}{\partial t} - A(t)\right]E = I \quad ; \quad E\left[\frac{\partial}{\partial t} - A(t)\right] = I - P_0 \ .$$

III. APPLICATION TO BOUNDARY VALUE PROBLEMS FOR ELLIPTIC EQUATIONS

From now on we suppose H finite-dimensional. Moreover the order of the operator A(t) of Ch. II will be one; and the letter m will be used to denote the degree of the linear partial differential operator in $X \times [0,T[$ under study, namely

$$\text{(III.1)} \qquad P = I\partial_t^m + \sum_{j=1}^{m} P_j(x,t,D_x)\,\partial_t^{m-j} \ ,$$

where, for each j = 1,..., m, $P_j(x,t,D_x)$ is a linear partial differential operator of order j in X whose coefficients are C^∞ functions of (x,t) in $X \times [0,T[$ valued in L(H) (thus P is a determined system of linear PDO). Actually we do not lose anything by taking the P_j to be classical pseudo-differential operators of order j respectively, in X, with values in L(H) - which we then denote by $P_j(t)$. For each j the principal symbol of $P_j(t)$ can be regarded as a C^∞ function of $((x,\xi),t)$ in $(T^*X \setminus 0) \times [0,T[$ valued in L(H), positive-homogeneous of degree j with respect to the fibre variable ξ . We are now going to make a truly restrictive hypothesis about our operator P, namely that its principal part is scalar; this will cover the main cases we shall be interested in:

(III.2) For every $j = 1,\dots, m$, the principal symbol $P_{j,0}(x,t,\xi)$ is equal to $P_j^0(x,t,\xi)I$, where $P_j^0(x,t,\xi)$ is complex valued.

Let us then select a scalar pseudodifferential operator of order j in X, depending smoothly on $t \in [0,T[$, with principal symbol $P_j^0(x,t,\xi)$, moreover classical; let it be P_j^0 $(1 \leq j \leq m)$ and set

(III.3) $$P = I\, P^0 + P' , \qquad P^0 = \partial_t^m + \sum_{j=1}^{m} P_j^0(t)\,\partial_t^{m-j}$$

where P' is not necessarily scalar, but has order $\leq m - 1$ (with respect to both x and t). Of course, if P is a differential operator, we may take each $P_j^0(t)$ to be one. Note that

(III.4) $$P' = \sum_{j=1}^{m} P_j'(t)\,\partial_t^{m-j} , \qquad \deg P_j'(t) \leq j - 1,\ 1 \leq j \leq m .$$

We now state the ellipticity hypothesis; it concerns the principal symbol of P, or rather that of P^0 :

(III.5) $$\sigma(P^0) = z^m + \sum_{j=1}^{m} P_j^0(x,t,\xi) z^{m-j} .$$

(III.6) There are two integers, m^+ and m^-, such that $m^+ + m^- = m$ and $m^- \geq 1$, such that, whatever $(x,\xi) \in T^*X \setminus 0$, $t \in [0,T[$, the polynomial $\sigma(P^0)$ with respect to z has exactly m^+ roots with real part > 0 and m- with real part < 0 .

Remark III.1.- When P is a differential operator or, more generally, an antipodal pseudodifferential operator, which means that $\sigma(P^0)(x,t,\xi;z) = \pm\, \sigma(P^0)(x,t,-\xi;-z)$, and provided that $\dim X > 1$, we must necessarily have $m^+ = m^- = m/2$ (and then m is necessarily even). This follows by a standard continuity argument and the fact that the complement of the origin in ξ-space is connected.

We shall use the ellipticity hypothesis (III.6) to factorize the principal symbol of P^0 as follows:

(III.7) $\sigma(P^0) = \sigma(M^{+0})\sigma(M^{-0}), \quad \sigma(M^{\pm 0}) = \prod_{k=1}^{m^{\pm}} \left[z - z_k^{\pm}(x,t,\xi)\right]$,

where the z_k^+ (resp., the z_k^-) are the roots with real part > 0 (resp., < 0). We wish to show that the factorization (III.7) leads to a factorization of the operator P itself, of the kind

(III.8) $P \sim M^+M^-$,

with $\sigma(M^{\pm}) = I\sigma(M^{\pm 0})$. If $m^+ = 0$ this is obvious : we may take $M^+ = I$. Thus, in the remainder, we shall assume $m^+ \geqslant 1$. Let us write

(III.9) $\sigma(M^{\pm 0})(x,t,\xi,z) = z^{m^{\pm}} + \sum_{j=1,\ldots,m^{\pm}} M_j^{\pm 0}(x,t,\xi)\, z^{m^{\pm}-j}$.

Although the roots $z_k^{\pm}$ cannot, in general, be represented as continuous functions of (x,t,ξ), for each j the coefficient $M_j^{+0}(x,t,\xi)$ (resp., $M_j^{-0}(x,t,\xi)$) is a C^{∞} function in $(T^*X \setminus 0) \times [0,T[$, positive-homogeneous of degree j with respect to ξ. This follows from the fact that these coefficients are symmetric functions of the roots z_k^+ (resp., z_k^-), and that the two sets of roots z_k^+ and z_k^- stay apart, as (x,t,ξ) varies.

Now, for each j, we may select a classical pseudodifferential operator in X, scalar, depending smoothly on $t \in [0,T[$, $M_j^{\pm 0}(t)$, with principal symbol $M_j^{\pm 0}(x,t,\xi)$, and we set

(III.10) $M^{\pm 0} = \partial_t^{m^{\pm}} + \sum_{j=1}^{m^{\pm}} M_j^{\pm 0}(t)\, \partial_t^{m^{\pm}-j}$,

we have

(III.11) $P = I\, M^{+0}M^{-0} + P''$,

with P" of the same type as P' in (III.3).

We shall reason by induction. Let us assume that we have found, for $N \geqslant 1$, operators $M^{\pm}_{(N)}$ and R_N such that

(III.12) $$M^{\pm}_{(N)} = \partial_t^{m^{\pm}} + \sum_{j=1,\dots,m^{\pm}} M^{\pm}_{(N)j}(t)\,\partial_t^{m^{\pm}-j} ,$$

(III.13) $\sigma(M^{\pm}_{(N)j}(t)) = I\, M^{\pm 0}_{j}(x,t,\xi)$ $(j = 1,\dots, m^{\pm})$ and the $M^{\pm}_{(N)j}(t)$ are classical pseudodifferential operators in X, valued in L(H), depending smoothly on t in $[0,T[$;

(III.14) $$R_{(N)} = \sum_{j=1,\dots,m} R_{(N)j}(t)\,\partial_t^{m-j} ;$$

(III.15) $$R_{(N)j}(t) \in C^{\infty}([0,T[;\ \Psi^{j-N}(X;L(H))),\ j = 1,\dots, m;$$

(III.16) $$P = M^{+}_{(N)}M^{-}_{(N)} + R_{(N)} .$$

We shall obtain all this with N + 1 in the place of N. In order to do this we regard $A^{\pm}_N = M^{\pm}_{(N+1)} - M^{\pm}_{(N)}$ as unknowns and write that

(III.17) $$R_{(N+1)} = M^{+}_{(N)}A^{-}_N + A^{+}_N M^{-}_{(N)} + A^{+}_N A^{-}_N - R_{(N)} .$$

We seek $A^{\pm}_N$ in the form $A^{\pm}_N = \sum_{j=1,\dots,m^{\pm}} A^{\pm}_{N,j}(t)\,\partial_t^{m^{\pm}-j}$ with $A^{\pm}_{N,j}(t)$ a classical pseudodifferential operator of order j-N in X, valued in L(H), depending smoothly on $t \in [0,T[$. We proceed as follows: We write that the principal symbol of the right-hand side in (III.17), regarded as an operator of order m - N, vanishes identically. Observe that the order of $A^{+}_N A^{-}_N$ is $\leq m^{+} + m^{-} - 2N < m - N$. Consequently we require that

(III.18) $$\sigma(M^{+}_{(N)})\,\sigma(A^{-}_N) + \sigma(M^{-}_{(N)})\,\sigma(A^{+}_N) = \sigma(R_{(N)}) ,$$

that is to say, by virtue of (III.13),

(III.19) $$\sigma(M^{+0})\,\sigma(A^{-}_N) + \sigma(M^{-0})\,\sigma(A^{+}_N) = \sigma(R_{(N)}) .$$

By their definition the polynomials $\sigma(M^{\pm 0})(x,t,\xi,z)$ (with respect to z) are <u>coprime</u>, and uniformly so as (x,ξ,t) remains in compact subsets of $(T^*X\setminus 0)\times[0,T[$ (observe that Eq. (III.19) is homogeneous with respect to (ξ,z)). By the classical Bezout's theorem, $\sigma(A_N^+)$ and $\sigma(A_N^-)$ are uniquely determined polynomials in z, of degrees $\leq m^+-1$, m^--1 respectively, with coefficients which are smooth functions of (x,ξ,t) (the degree of homogeneity of these symbols with respect to (ξ,z) is m^+-1 and m^--1 respectively). It only remains to select for $A^{\pm}_{N,j}(t)$ classical pseudodifferential operators in X, valued in L(H), depending smoothly on t, with the appropriate principal symbols (specifically $\frac{1}{j!}\frac{\partial^j}{\partial z^j}\sigma(A_N^{\pm})\big|_{z=0}$): if (III.17) is taken as definition of $R_{(N+1)}$ we shall have all the properties from (III.12) to (III.16) satisfied with N+1 instead of N. Since we know, by (III.11), that they are satisfied when N = 1, we can go to the limit for N arbitrarily large, namely $N=+\infty$ (indeed, the operators $M^{\pm}_{(N)}$ obviously converge). At the limit, rather than writing $N=+\infty$, we shall omit the subscripts N. Thus (III.8) is satisfied. Actually we shall go one step further and divide R by M^-:

$$\text{(III.20)}\qquad R = Q\,M^- + R', \qquad R' = \sum_{j=1,\dots,m^-} R'_j(t)\,\partial_t^{m^--j}\ ;$$

$$\text{(III.21)}\qquad R'_j(t) \in C^\infty([0,T[;\Psi^{-\infty}(X;L(H))),\ j=1,\dots,m^-.$$

This poses no problem: it is just a question of replacing a large enough number of times $\partial_t^{m^-}$ by $M^- - \sum_{j=1,\dots,m^-} M_j^-(t)\,\partial_t^{m^--j}$. Finally (after dropping the primes ' in the notation for R' and substituting M^+ for $M^+ + Q$) we obtain the decomposition

$$\text{(III.22)}\qquad P = M^+ M^- + R\ ,$$

with $M^{\pm}$ of the same kind as $M^{\pm 0}$ (see (III.10)) and R the same as R' in (III.20)-(III.21).

Now suppose that we are dealing with the equation

(III.23) $Pu = f, \quad u, f \in C^\infty([0,T[; \mathcal{D}'(X;H)).$

By (III.22) we see that it is equivalent to the system of (two) equations

(III.24) $M^- u = v$,

(III.25) $M^+ v + Ru = f$.

Let us consider a typical equation of the kind

(III.26) $Mw = g, \quad M = \partial_t^r + \sum_{j=1,\dots,r} C_j(t)\partial_t^{r-j}$,

where, for each j, $C_j(t)$ is a pseudodifferential operator in X of order j with values in L(H), depending smoothly on t. Let us select once and for all an elliptic pseudodifferential operator Λ of order <u>one</u> in X, <u>scalar</u>; we take Λ to be <u>classical</u>, and <u>properly supported</u>. We shall furthermore require that there be a (classical elliptic) pseudodifferential operator of order - 1 in X, properly supported, which we denote by Λ^{-1} and such that $\Lambda\,\Lambda^{-1}$ = Identity of $\mathcal{D}'(X)$. Let us not worry whether such an operator does exist: When the manifold X is <u>compact</u>, which is the only case that truly interests us, it certainly does: Equip X with a Riemanniam metric, let Δ_X denote the Laplace-Beltrami operator on X for that metric (we shall suppose here that Δ_X is ≤ 0) ; a possible choice is $\Lambda = (1 - \Delta_X)^{\frac{1}{2}}$. (When X is compact, the requirement "properly supported" is void, and $\mathcal{D}'(X;H) = \mathcal{E}'(X;H)$.)

Let us then set, for each j = 1,..., r, $w^j = \Lambda^{1-j}\partial_t^{j-1}w$; we shall denote by $\vec{W}$ the r-vector with components $w^1,\dots,w^r$. Each component is a smooth function of t with values in $\mathcal{D}'(X;H)$, hence $\vec{W}$ itself is such a function but valued in $\mathcal{D}'(X;H\otimes\mathbb{C}^r)$. Note that we have:

(III.27) $\partial_t w^j = \Lambda w^{j+1}$, $\quad j = 1,\dots, r-1$.

If we multiply both members in Eq. (III.26), by Λ^{1-r}, we may rewrite it in the following manner:

$$\partial_t w^r + \sum_{j=1}^{r} \Lambda^{1-r} c_j(t) \Lambda^{r-j} w^{r-j+1} = \Lambda^{1-r} g . \tag{III.28}$$

We observe that the "coefficients" $c_j^{\#}(t) = \Lambda^{1-r} c_{r-j+1}(t) \Lambda^{j-1}$ are pseudodifferential operators of order <u>one</u> in X, valued in L(H), depending smoothly on t. We gather the equations (III.27) & (III.28) together in a single system,

$$\partial_t \vec{W} - \mathcal{M}(t)\vec{W} = \vec{G} , \tag{III.29}$$

with $\vec{G}$ the r-vector with components all zero, except the r-th one, equal to $\Lambda^{1-r} g$, and where $\mathcal{M}(t)$ is the $r \times r$ matrix

$$\begin{pmatrix} 0 & I\Lambda & 0 & & 0 \\ 0 & 0 & I\Lambda & \cdots & 0 \\ \cdots & \cdots & \cdots & \cdots & \cdots \\ 0 & 0 & 0 & \cdots & I\Lambda \\ -c_1^{\#}(t) & -c_2^{\#}(t) & -c_3^{\#}(t) & \cdots & -c_r^{\#}(t) \end{pmatrix} . \tag{III.30}$$

Of course we view $\mathcal{M}(t)$ as pseudodifferential operator of order <u>one</u> in X with values in $L(H \otimes \mathbb{C}^r)$. A standard and important remark is that its principal symbol $\sigma(\mathcal{M}(t))$ is such that

$$\det\{zI - \sigma[\mathcal{M}(t)]\} = \sigma(M)(x,t,\xi,z) , \tag{III.31}$$

where we view $\sigma(\mathcal{M}(t))$ as a matrix over the <u>ring</u> L(H), depending smoothly on $((x,\xi),t) \in (T^*X \setminus 0) \times [0,T[$, and positive-homogeneous of degree one with respect to ξ. Thus the determinant det is computed in that ring (although the ring L(H) is not commutative if $\dim H > 1$, the computation of that determinant is made easy by the fact that all rows, except possibly the last one, in $\sigma(\mathcal{M}(t))$, are scalar multiples of the identity of H ; at any rate, in the application of what precedes to Eqq. (III.24) and (III.25) also the last row will be a scalar multiple of the identity).

First we apply the preceding transformation to Eq. (III.24); in this case $r = m^-$; we set $u^j = \Lambda^{1-j}\partial_t^{j-1}u$, $j = 1,\dots, m^-$, and shall denote by $\vec{u}$ the m^--vector with components u^j. Let us set right away $v^j = \Lambda^{1-j}\partial_t^{j-1}v$, $j = 1,\dots, m^+$, and call $\vec{v}$ the vector with components v^j. We then denote by Jv the m^--vector whose components are all zero, except the last one, equal to $\Lambda^{1-m^-}v$. With this notation Eq. (III.24) reads

(III.32) $$\partial_t\vec{u} - A^-(t)\vec{u} = J\vec{v}\ .$$

According to (III.31) we have (cf. (III.10) & (III.22)):

(III.33) $$\det\left\{zI - \sigma\left[A^-(t)\right]\right\} = \sigma(M^{-0})(x,t,\xi,z)\ .$$

where the determinant is now computed in the complex field (we have taken advantage of the fact that the principal symbols of the operators $M_j^-(t)$ are scalar multiples of the identity of H).

On the other hand we note that, according to (III.20),

(III.34) $$\Lambda^{1-m^+}Ru = \sum_{j=1,\dots,m^-} \Lambda^{1-m^+}R_j(t)\Lambda^{m^--j}u^{m^--j+1}\ .$$

We may then denote by $\mathcal{R}\vec{u}$ the m^+-vector with components all zero, except the last one, equal to $\Lambda^{1-m^+}Ru$. It is clear that

(III.35) $\mathcal{R}$ <u>is a regularizing operator in</u> X, <u>depending smoothly on</u> $t \in [0,T[$, <u>valued in</u> $L(H\otimes\mathbb{C}^{m^-};H\otimes\mathbb{C}^{m^+})$.

We now denote by $\vec{g}$ the m^+-vector with all components equal to zero, except the last one, equal to $\Lambda^{1-m^+}f$. Thus Eq. (III.25) reads:

(III.36) $$\partial_t\vec{v} - A^+(t)\vec{v} = \vec{g} - \mathcal{R}\vec{u}\ .$$

We have:

(III.37) $$\det\{zI - \sigma[A^{+}(t)]\} = \sigma(M^{+0})(x,t,\xi,z) \ .$$

The relations (III.36) and (III.37) show that the eigenvalues of $A^{-}(t)$ and those of $- A^{+}(t)$ stay in the open half-plane C_{-} (they are roots z_k^{-}, and the opposites $- z_k^{+}$, of the polynomial $\sigma(P^0)(x,t,\xi,z)$). It ought to be underlined that we are viewing, here, $\sigma(A^{\pm}(t))$ as matrices of size $m^{\pm} \times m^{\pm}$. To have them as symbols valued in the space of linear mappings $H \otimes \mathbf{C}^{m^{\pm}} \longrightarrow H \otimes \mathbf{C}^{m^{\pm}}$, one must then tensor them (on the left) with the identity of H.

We may thus state that

(III.38) <u>the basic hypothesis</u>, (II.5), <u>of Ch. II</u>, <u>is verified by</u> $A^{-}(t)$ <u>and by</u> $- A^{+}(t)$.

The next step is to adjoin "initial conditions" to Eq. (III.23):

(III.39) $$B_j(x,D_x,\partial_t)u\big|_{t=0} = h_j \quad (j = 1,\dots,\nu).$$

Here $h_j \in \mathcal{D}'(X;H)$ for each j, and

(III.40) $$B_j(x,D_x,\partial_t) = \sum_{k=0,\dots,d_j} B_{j,k}(x,D_x)\,\partial_t^k \ ,$$

where, for each choice of $j = 1,\dots,\nu$, $k = 1,\dots,d_j$, $B_{j,k}(x,D_x)$ is a pseudodifferential operator in X, valued in L(H).

The first thing we do is to divide each $B_j = B_j(x,D_x,\partial_t)$ by P, using the fact that P is a monic polynomial with respect to ∂_t:

(III.41) $$B_j = Q_j'P + B_j' \ .$$

Note that the degree of B_j' as a polynomial in ∂_t does not exceed m - 1. We then replace the conditions (III.39) by

(III.42) $$B_j'u\big|_{t=0} = h_j - (Q_j'f)\big|_{t=0} \ .$$

Next we divide B'_j by M^- :

(III.43) $$B'_j = Q_j M^- + B^\#_j \qquad (j = 1,\dots,\nu).$$

Here not only is the degree of $B^\#_j \leq m^- - 1$ (as polynomial in ∂_t) but also now $\deg Q_j \leq \deg B'_j - \deg M^- \leq m - m^- - 1 = m^+ - 1$. By virtue of (III.24) we may replace the conditions (III.42) by

(III.44) $$B^\#_j u\big|_{t=0} = h_j - (Q'_j f)\big|_{t=0} - (Q_j v)\big|_{t=0} .$$

Let us denote by $\vec{h}_\#$ the ν-vector with components $h_j - (Q'_j f)\big|_{t=0}$ ($j = 1,\dots,\nu$), by $\mathfrak{Q}\vec{v}(0)$ the one whose components are $(Q_j v)\big|_{t=0}$. The fact that degree (in ∂_t) of Q_j is $\leq m^+ - 1$ implies that $\mathfrak{Q}$ may indeed be regarded as a $\nu \times m^+$ matrix, since

(III.45) $$Q_j v = \sum_{k=0,\dots,m^+-1} Q_{j,k}(t)\,\partial_t^k v = \sum_{k=1}^{m^+} Q_{j,k-1}(t)\Lambda^{k-1} v^k .$$

Finally we may rewrite the initial conditions (III.39) in the manner

(III.46) $$\mathfrak{B}\,\vec{u}(0) = \vec{h}_\# - \mathfrak{Q}\,\vec{v}(0) ,$$

where $\mathfrak{B}$ is the pseudodifferential operator in X, valued in the space of $\nu \times m^-$ matrices with entries in L(H) defined as follows: If one writes

(III.47) $$B^\#_j = \sum_{k=0}^{m^--1} B^\#_{j,k}(t)\,\partial_t^k ,$$

we have

(III.48) $$B^\#_j u = \sum_{k=1}^{m^-} B^\#_{j,k-1}(t)\Lambda^{k-1} u^k .$$

Consequently, if $\mathfrak{B}_{j,k}$ ($j=1,\dots,\nu$, $k=1,\dots,m^-$) is a generic entry of $\mathfrak{B}$ we have

(III.49) $$\mathfrak{B}_{j,k} = B^\#_{j,k-1}(0)\Lambda^{k-1} .$$

Let us summarize what we have done so far in this chapter:

The system of equations (III.23), (III.39) :

$$(*) \qquad Pu = f, \qquad B_j(x,D_x,\partial_t)u\big|_{t=0} = h_j \quad (1\leq j\leq \nu) ,$$

has been transformed into the system (III.32), (III.36), (III.46):

$$(**) \qquad \partial_t\vec{u} - A^-(t)\vec{u} = J\vec{v}, \qquad \mathcal{B}\,\vec{u}(0) = \vec{h}_\# - \mathcal{Q}\,\vec{v}(0) ;$$

$$(***) \qquad \partial_t\vec{v} - A^+(t)\vec{v} = \vec{g} - \mathcal{R}\,\vec{u} .$$

Let us emphasize the fact that these equations are exact (in so far as $\Lambda\,\Lambda^{-1} = I$ exactly, which is possible to achieve when X is compact). Nevertheless we have not quite succeeded in "triangulazing" the problem (*) : we have only approximately done so, since Eq. (***) still contains $\mathcal{R}\vec{u}$. But we shall see that the above transformation still enables us to analyze some important aspects of (*).

First of all let us point out that there is a one-to-one correspondence between solutions of (*) and solutions of (**)-(***): the argument in the preceding pages has shown how to go from $u \in C^\infty([0,T[;\mathcal{D}'(X;H))$ to $\vec{u} \in C^\infty([0,T[;\mathcal{D}'(X;H\otimes\mathbb{C}^{m^-}))$, $\vec{v} \in C^\infty([0,T[;\mathcal{D}'(X;H\otimes\mathbb{C}^{m^+}))$. Conversely one can go from $\vec{u}$ to u simply by taking the latter to be the <u>first</u> component of the former.

However, it might be as difficult to solve (**)-(***) exactly as it is to solve (directly and exactly) (*). We shall replace (***) by the following approximation:

$$(\text{III.50}) \qquad \partial_t\vec{v}_\# - A^+(t)\vec{v}_\# = \vec{g} \qquad \underline{\text{in}}\ X\times[0,T[,$$

and modify also (**) accordingly:

$$(\text{III.51}) \qquad \partial_t\vec{u}_\# - A^-(t)\vec{u}_\# = J\vec{v}_\# \qquad \underline{\text{in}}\ X\times[0,T[,$$

$$(\text{III.52}) \qquad \mathcal{B}\,\vec{u}_\#(0) = \vec{h}_\# - \mathcal{Q}\,\vec{v}_\#(0) \quad \underline{\text{in}}\ X.$$

This modified problem we can easily solve - provided we strengthen a little bit our hypotheses on $A^{\pm}(t)$ and on f, and by way of consequence on $\vec{g}$: specifically that we assume that all these functions are smooth, with respect to t, in the closed interval $[0,T]$. This is of no great import in the applications (it can be achieved by slightly decreasing T).

Since $-A^{+}(t)$ satisfies the basic assumption (II.5) we can solve the backward Cauchy problem for Eq. (III.50), starting at t = T, and thus write (for an arbitrary choice of $\vec{v}_{\#}(T) \in \mathcal{D}'(X;H)$)

$$\text{(III.53)} \qquad \vec{v}_{\#}(t) = U^{+}(t,T)\vec{v}_{\#}(T) - \int_t^T U^{+}(t,t')\vec{g}(t')dt' ,$$

where $U^{+}(t,t')$ is the relevant parametrix. We can then put the solution $\vec{v}_{\#}$ of (III.50) thus obtained into (III.51)-(III.52) and solve the forward Cauchy problem, starting at t = 0 (cf. (II.45)):

$$\text{(III.54)} \qquad \vec{u}_{\#}(t) = U^{-}(t,0)\vec{u}_{\#}(0) + \int_0^t U^{-}(t,t')\, J\vec{v}_{\#}(t')dt' ,$$

where $U^{-}(t,t')$ is the parametrix for (III.51)-(III.52). In doing this we have taken advantage of the fact that $A^{-}(t)$ satisfies (II.5).

Of course, once we use the parametrices for these various Cauchy problems, we only obtain approximate solutions - modulo errors that involve regularizing operators acting on the various data. The question is then to go from these solutions to solutions of (**) & (***), and from there to solutions of our original problem (*). Here we shall focus on the regularity of these solutions, and show how the regularity of one set of solutions determines that of the other.

Lemma III.1.- Let $\vec{u}$, $\vec{v}$ be solutions of (**), (***) where we assume that g belongs to $C^{\infty}([0,T];\mathcal{D}'(X;H\otimes\mathbb{C}^{m^{+}}))$, and let $\vec{u}_{\#}$, $\vec{v}_{\#}$ be defined by (III.53)-(III.54). Then

$$\text{(III.55)} \qquad \vec{v} - \vec{v}_{\#} \in C^{\infty}(X \times [0,T[\,;\, H \otimes \mathbb{C}^{m^{+}}) .$$

If moreover we assume that $\vec{u}_{\#}(0) - \vec{u}(0) \in C^{\infty}(X;H\otimes \mathbb{C}^{m^-})$, then

(III.56) $$\vec{u} - \vec{u}_{\#} \in C^{\infty}(X\times[0,T[;H\otimes \mathbb{C}^{m^-}) .$$

Proof: It is very simple: it suffices to observe that $\vec{u}_1 = \vec{u} - \vec{u}_{\#}$, $\vec{v}_1 = \vec{v} - \vec{v}_{\#}$ satisfy the equations

(III.57) $$\left[\partial_t - A^-(t)\right]\vec{u}_1 = J\vec{v}_1 + \vec{\phi} \quad ,$$

(III.58) $$\left[\partial_t - A^+(t)\right]\vec{v}_1 = \vec{\psi} \ ,$$

where $\vec{\phi}$ and $\vec{\psi}$ are linear combinations of $\vec{u}$, $\vec{v}$, $\vec{u}_1$, $\vec{v}_1$ with coefficients which are matrices, of the appropriate size, with regularizing operators as entries. We apply then Formula (II.45) to $\vec{u}_1$, and the analogous Eq. (III.58) to $\vec{v}_1$:

(III.59) $$\vec{u}_1(t) = U^-(t,0)\vec{u}_1(0) + \int_0^t U^-(t,t')\left[J\vec{v}_1(t') + \vec{\phi}(t')\right] dt' - R_0^-(t)\vec{u}_1 + \int_0^t R^-(t,t')\vec{u}_1(t')dt' ,$$

(III.60) $$\vec{v}_1(t) = U^+(t,T)\vec{v}_1(T) - \int_t^T U^+(t,t')\,\vec{\psi}(t')dt' - R_0^+(t)\vec{v}_1(t) + \int_t^T R^+(t,t')\vec{v}_1(t')dt' ,$$

where the various R's stand for regularizing operators, depending smoothly on t or on (t,t'). First (III.55) follows at once from (III.60) if we recall that $U^+(t,T)$ is regularizing as soon as $t < T$. Taking this fact into account in (III.59) we deduce (III.56) if we assume that $\vec{u}_1(0)$ is C^{∞} in X.

Q. E. D.

Definition III.1.- The problem (**)-(***) is said to be hypo-elliptic if given any open subset Y of X, any data $\vec{g} \in C^{\infty}([0,T];\mathcal{D}'(X;H\otimes \mathbb{C}^{m^+}))$, $\vec{h}_{\#}$ $\mathcal{D}'(X;H\otimes\mathbb{C}^{\nu})$ whose restrictions to Y are smooth, i. e.,

(III.61) $$\vec{g} \in C^{\infty}(Y\times[0,T];H\otimes\mathbb{C}^{m^+}) , \quad \vec{h}_{\#} \in C^{\infty}(Y;H\otimes \mathbb{C}^{\nu}),$$

every solution $(\vec{u},\vec{v})$ of (**)-(***), such that

(III.62) $\vec{u} \in C^\infty([0,T];\mathcal{D}'(X;H\otimes\mathbb{C}^{m^-})),\ \vec{v} \in C^\infty([0,T];\mathcal{D}'(X;H\otimes\mathbb{C}^{m^+})),$

is in fact smooth in Y for t T, i. e.,

(III.63) $\vec{u} \in C^\infty(Y;H\otimes\mathbb{C}^{m^-}),\quad \vec{v} \in C^\infty(Y;H\otimes\mathbb{C}^{m^+}).$

Remark III.2.- There would be no gain in generality in relaxing Condition (III.62) and allowing $\vec{u}$ and $\vec{v}$ to be distributions in t. Indeed, by a standard argument, our hypotheses on $\vec{g}$ and the equations (**)-(***) themselves would automatically imply (III.62) (one could allow g not be smooth with respect to t outside of Y, but what counts is its smoothness in Y).

Remark III.3.- The hypo-ellipticity of Problem (**)-(***) is obviously equivalent to that of our original problem (*) (defined in evident manner).

Finally we recall that the operator $\mathcal{B}$ in X is said to be hypo-elliptic if it preserves the singular supports.

Remark III.4.- If one prefers to consider "wave-front sets" hypo-ellipticity (cf. Def. I.8), namely the property that the operator under study preserves the wave-front sets, and not merely the singular supports, it should be said that all the statements in this last part of Ch. III, in particular Th. III.1 below, have their counterparts valid for this concept.

Theorem III.1.- The problem (**)-(***) is hypo-elliptic if and only if the pseudodifferential operator $\mathcal{B}$ in X is hypo-elliptic.

Proof: Suppose first that $\mathcal{B}$ is hypo-elliptic. By the fact that pseudodifferential operators, here $U^+(t,t')$, are pseudolocal, we derive from (III.54) that $\vec{v}_\#$ is smooth in $Y \times [0,T[$ (remember that $U^+(t,T)$ is regularizing for $t < T$). By the first part of Lemma III.1 we conclude that also $\vec{v}$ is smooth in $Y\times[0,T[$; in particular $\vec{v}(0) \in C^\infty(Y;H\otimes\mathbb{C}^{m^+})$. We take this into account in the relation $\mathcal{B}\,\vec{u}(0) = \vec{h}_{\#} - \mathcal{Q}\,\vec{v}(0)$, and derive that $\vec{u}(0) \in C^{\infty}(Y;H \otimes \mathbb{C}^{m^-})$. It suffices then to apply the last part of Lemma III.1 with $\bar{u}_{\#}(0) = \bar{u}(0)$.

Suppose now that $\mathcal{B}$ is not hypo-elliptic: there is then a distribution $\vec{u}_o$ in X, valued in $H \otimes \mathbb{C}^{m^-}$, whose restriction to some open set $Y \subset X$ is not C^{∞} but such that the one of $\mathcal{B}\,\vec{u}_o$ is C^{∞}. Set $\vec{w}(t) = U^-(t,0)\vec{u}_o$; then:

(III.64) $\quad \bar{v}_* = \left[\frac{\partial}{\partial t} - A^-(t)\right]\vec{w}$ is a C^{∞} function $X \times [0,T] \to H \otimes \mathbb{C}^{m^-}$.

Note that $v_*^j = \partial_t w^j - \Lambda w^{j+1}$ if $j < m^-$. Let us define inductively

(III.65) $\quad u^{m^-} \doteq w^{m^-}$,

(III.66) $\quad u^j = w^j + \int_0^t \left[\Lambda(u^{j+1} - w^{j+1})(t') - v_*^j(t')\right] dt'$ if $j < m^-$.

By (descending) induction on j and by (III.64) we see that $\vec{u} - \vec{w}$ is C^{∞} in $X \times [0,T]$ with values in $H \otimes \mathbb{C}^{m^-}$. We have:

(III.67) $\quad \partial_t u^j = \Lambda u^{j+1}$ if $j < m^-$.

Let then χ denote the last (i. e., the m^-th) component of $\partial_t u - A^-(t)u$, and define $v = \Lambda^{m^- -1}\chi$, and the m^+-vector $\vec{v} = (v^1,\dots,v^{m^+})$ as before, by setting $v^j = \Lambda^{1-j}\partial_t^{j-1}v$. Eq. (III.32) is automatically satisfied. Since

$$\left[\partial_t - A^-(t)\right]\vec{u} = \vec{v}_* + \left[\partial_t - A^-(t)\right](\vec{u} - \vec{w}) ,$$

we see that $\vec{v} \in C^{\infty}(X \times [0,T];H \otimes \mathbb{C}^{m^+})$. We then set $\vec{g} = \left[\partial_t - A^+(t)\right]\vec{v} + \mathcal{R}\,\vec{u}$; we have $\vec{g} \in C^{\infty}(X \times [0,T];H \otimes \mathbb{C}^{m^+})$. On the other hand, $\vec{u}(0) = \vec{w}(0) = \vec{u}_o$. By virtue of our hypothesis about $\mathcal{B}\,\vec{u}_o$, $\vec{h}_{\#} = \mathcal{B}\,\vec{u}_o + \mathcal{Q}\,\vec{v}(0)$ is a C^{∞} mapping $Y \to H \otimes \mathbb{C}^{\nu}$. Thus (III.61) and, of course, (III.62) are true, but (III.63) is not.

Q. E. D.

IV. COERCIVE BOUNDARY VALUE PROBLEMS

Coercive problems are a particular case of the type of problems (*): In the present chapter P will be the same elliptic operator as in Ch. III. But for the sake of simplicity we shall assume that dim H = + 1, in other words the operator P will be scalar.

Coercivity is characterize by two conditions:

(IV.1) ν, the number of boundary conditions, is equal to m^-, the number of roots of the polynomial in z, $\sigma(P)(x,t,\xi,z)$, with real part < 0 ;

(IV.2) the principal symbols $\sigma(B_j)(x,\xi,z)$ of the boundary operators $B_j(x,D_x,\partial_t)$ $(j = 1,\dots,\nu)$ are linearly independent modulo $\sigma(M^{-0})(x,0,\xi,z)$ (see (III.7)), whatever (x,ξ) in $T^*\wedge \setminus 0$.

It should be noted that the principal symbols $\sigma(B_j)(x,\xi,z)$ are positive homogeneous functions of (ξ,z) of respective degrees m_j (with no relation, a priori, to their degrees, d_j, as polynomials in z). Actually we may multiply each $B_j(x,D_x,\partial_t)$ by Λ^{-m_j} and assume henceforth that their homogeneity degree with respect to (ξ,z) is equal to zero - whatever j. This is what we shall do.

Since, by (IV.1), $\nu = \deg_z \sigma(M^{-0})$, we see that the $\sigma(B_j)$ form a linear basis of the space of polynomials in the variable z modulo $\sigma(M^{-0}(0))$. We may effect the division of $\sigma(B_j)(x,\xi,z)$ by $\sigma(M^{-0})(x,0,\xi,z)$ (as polynomials in z):

(IV.3) $$\sigma(B_j)(x,\xi,z) = Q_j''(x,\xi,z)\,\sigma(M^{-0})(x,0,\xi,z) + b_j(x,\xi,z)\ ,$$

with $\deg_z b_j \leq m^- - 1$. The polynomials b_j form a basis of the vector space of polynomials in z of degree $< m^-$.

Now, if we go back to the division formulas (III.41) & (III.43), we see that, for each j, $b_j(x,\xi,z)$ is the principal symbol of $B_j^{\#}(0)$, that is to say, according to (III.47),

$$b_j(x,\xi,z) = \sum_{k=0}^{m^- -1} \sigma(B_{j,k}^{\#})(x,0,\xi)\, z^k \,. \tag{IV.4}$$

Because of the homogeneity of degree zero of b_j with respect to (ξ,z) we see that $\sigma(B_{j,k}^{\#})(x,0,\xi)$ is positive-homogeneous of degree $-k$ with respect to ξ, and therefore, if we set, as in (III.49), $\mathcal{B}_{j,k} = B_{j,k-1}^{\#}(0)\Lambda^{k-1}$, we obtain that $\sigma(\mathcal{B}_{j,k})(x,\xi)$ is positive-homogeneous of degree zero with respect to ξ. It is then clear that the coercivity condition (IV.2) is equivalent to the property that

(IV.5) <u>The vectors</u> $\beta_j(x,\xi) = (\sigma(\mathcal{B}_{j,k})(x,\xi))_{k=1,\dots,\nu}$ <u>are linearly independent, whatever</u> $(x,\xi) \in T^*X \setminus 0$,

or, in other words,

(IV.6) <u>the</u> $m^- \times m^-$ <u>matrix</u> $\sigma(\mathcal{B})(x,\xi) = (\sigma(\mathcal{B}_{j,k})(x,\xi))_{j,k=1,\dots,m^-}$ <u>is invertible, whatever</u> (x,ξ) <u>in</u> $T^*X \setminus 0$.

Since we can retrace our steps and clearly go back from (IV.6) to (IV.2) we may state:

<u>Proposition IV.1</u>.- <u>The boundary value problem</u> (*) <u>is coercive if and only if</u> $\nu = m^-$ <u>and the boundary operator</u> $\mathcal{B}$ <u>is elliptic</u> (with the present choice of the operators B_j, $\mathcal{B}$ is a pseudodifferential operator of order zero in X).

By combining this with Th. III.1 (and Cor. I.5 & remark following it) we obtain:

<u>Theorem IV.1</u>.- <u>If the boundary value problem</u> (*) <u>is coercive, it is hypo-elliptic</u>.

The coercive estimates

In this section we indicate briefly how to derive the standard coercive estimates from Th. II.4. We assume throughout that the pseudodifferential operators B_j $(j=1,\dots,m^-)$ and $\mathcal{B}$ are of order zero ($\mathcal{B}$ is elliptic). We continue to reason under the same hypotheses as in the first part of this chapter, in particular we shall assume that all data, $A^{\pm}(t)$, $\bar{g}(t)$ are C^∞ in the closed interval $[0,T]$. We shall use Sobolev norms on distributions in X without distinguishing where the values of those distributions lie.

We shall consider an arbitrary function belonging to $C_c^\infty(X\times[0,T[)$ (thus it vanishes for $t > T_0$ where T_0 is some number $< T$) which we denote by u, and define $f = Pu$, $u_j = B_j(x,D_x,\partial_t)u|_{t=0}$ $(j = 1,\dots,m^-)$. We also define the vector-valued functions $\vec{u}$, $\vec{v}$ as indicated in Ch. III. In what follows C will be a constant independent of u (and therefore of $T_0 < T$). We begin by applying (II.61) (keeping in mind that what was denoted by m in Ch. II is now equal to one, whereas m denotes now the degree of P):

$$\text{(IV.7)}\qquad \int_0^T \|\partial_t^k\vec{u}(t)\|_s^2\,dt \leq C\Big\{\|\vec{u}(0)\|^2_{s+k-\frac{1}{2}} + \int_0^T \|\vec{u}(t)\|^2_{s'}\,dt + \sum_{j\leq\sup(k-1,0)} \int_0^T \|\partial_t^j\, J\vec{v}(t)\|^2_{s+k-j-1}\,dt\Big\}.$$

Actually, when $k = 0$, (IV.7) follows from (II.60). We have used the fact that $J\vec{v} = \partial_t\vec{u} - A^-(t)\vec{u}$. Now, using the fact that $\mathcal{B}$ is elliptic of order zero, we get

$$\text{(IV.8)}\qquad \|\vec{u}(0)\|_s^2 \leq C\{\|\mathcal{B}\vec{u}(0)\|_s^2 + \|\vec{u}(0)\|^2_{s'}\}.$$

On the other hand,

$$\text{(IV.9)}\qquad \|\vec{u}(0)\|^2_{s'} = -2\,\mathrm{Re}\int_0^T\{(\vec{u},A^-(t)\vec{u})_{s'} + (\vec{u},\partial_t\vec{u} - A^-(t)\vec{u})_{s'}\}\,dt$$
$$\leq C\int_0^T\Big(\|\vec{u}(t)\|^2_{s'+\frac{1}{2}} + \|[\partial_t - A^-(t)]\vec{u}(t)\|^2_{s'-\frac{1}{2}}\Big)dt.$$

We apply (III.46): $\vec{u}(0) = \vec{h}_{\#} - \mathcal{Q}\,\vec{v}(0)$. By (III.41) and (III.43) we see that B'_j and $B^{\#}_j$ are pseudodifferential operators in (x,t) of degree zero, like B_j . Therefore the Q_j are of order $-m^-$, and so is $\mathcal{Q}$. After a redefinition of s' we derive from this, and from (IV.8), (IV.9):

$$\text{(IV.10)}\qquad \|\vec{u}(0)\|^2_{s+k-\frac{1}{2}} \le C\left\{\|\vec{h}_{\#}\|^2_{s+k-\frac{1}{2}} + \|\vec{v}(0)\|^2_{s+k-m^--\frac{1}{2}} + \int_0^T\left(\|\vec{u}(t)\|^2_{s'} + \|\vec{v}(t)\|^2_{s'}\right)dt\right\} .$$

We apply the analogue of (IV.9) with $v(0)$ substituted for $u(0)$ and $s' = s + k - m^- - 1/2$. We get

$$\text{(IV.11)}\qquad \|\vec{v}(0)\|^2_{s+k-m^--\frac{1}{2}} \le C\int_0^T\left(\|\vec{v}(t)\|^2_{s+k-m^-} + \left\|\left[\partial_t - A^+(t)\right]\vec{v}(t)\right\|^2_{s+k-m^--1}\right)dt .$$

We may also apply the analogue of (IV.7) for $\vec{v}$ instead of $\vec{u}$, keeping in mind that we must then exchange 0 and T and that $\vec{v}(T) = 0$ (since $u^1(T) = 0$ and $v^1 = M^-u^1$):

$$\text{(IV.12)}\qquad \int_0^T\|\partial_t^{\ell}\vec{v}(t)\|^2_s\,dt \le C\left\{\int_0^T\|\vec{v}(t)\|^2_{s'}\,dt + \sum_{j\le\sup(\ell-1,0)}\int_0^T\left\|\partial_t^j\left\{\left[\partial_t - A^+(t)\right]\vec{v}(t)\right\}\right\|^2_{s+\ell-j-1}dt\right\}.$$

We first apply (IV.12) where we substitute $s + k - m^-$ for s in conjunction with (IV.11). We note that $\partial_t\vec{v} - A^+(t)\vec{v} = \vec{g} - \mathcal{R}\,\vec{u}$, and that all components of $\vec{g}$ are zero, except the last one, equal to $\Lambda^{1-m^+}f$. We obtain:

$$\text{(IV.13)}\qquad \|\vec{v}(0)\|^2_{s+k-m^--\frac{1}{2}} \le C\left\{\int_0^T\|f(t)\|^2_{s+k-m}\,dt + \int_0^T\left(\|\vec{u}(t)\|^2_{s'} + \|\vec{v}(t)\|^2_{s'}\right)dt\right\} .$$

We put this back into (IV.10) and from there into (IV.7), where we now

$s = m - k$ (and redefine s') :

$$\text{(IV.14)} \quad \int_0^T \|\partial_t^k \vec{u}(t)\|^2_{m-k}\, dt \leq C\Big\{\|\vec{h}_\#\|^2_{m-\frac{1}{2}} + \int_0^T \|f(t)\|_0^2\, dt + \sum_{j\leq \sup(k-1,0)} \int_0^T \|\partial_t^j v^1\|^2_{m^+-j}\, dt + \int_0^T \left(\|\vec{u}\|^2_{s'} + \|\vec{v}\|^2_{s'}\right) dt\Big\}.$$

We have used the fact that all components of $J\vec{v}$ are zero, except the last one, equal to $\Lambda^{1-m^-}v^1$, and also the fact that $m - m^- = m^+$. Now we use the property that, if $j < m^+$, $\partial_t^j v^1 = \Lambda^j v^{j+1}$. We obtain, provided that k remain $\leq m^+ + 1$, and $j \leq k - 1$,

$$\text{(IV.15)} \quad \int_0^T \|\partial_t^j v^1\|^2_{m^+-j}\, dt \leq C\int_0^T (\|\vec{v}\|^2_{m^+} + \|\partial_t \vec{v}\|^2_{m^+-1})\, dt .$$

We apply once again (IV.12) with $\ell = 0, 1$. We see that the right-hand side in (IV.15) does not exceed a constant times

$$\int_0^T \left(\|\vec{g}(t)\|^2_{m^+-1} + \|\vec{u}(t)\|^2_{s'} + \|\vec{v}(t)\|^2_{s'}\right) dt .$$

Finally, we recall the definition of $\vec{g}$, and the fact that

$$\text{(IV.16)} \quad \int_0^T \|\vec{v}\|^2_{s'}\, dt \leq C \sum_{k=0}^{m-1} \int_0^T \|\partial_t^k u^1\|^2_{s''}\, dt$$

for a suitable s'' (which can be as close as we wish to $-\infty$ if we choose $-s'$ large enough). We also use the fact that $\partial_t^k u^{j+1} = \Lambda^j \partial_t^{k+j} u^1$ and that if k varies from 0 to $m^+ + 1$, and j from 0 to $m^- - 1$, then $j + k$ varies from 0 to m. We obtain the <u>coercive estimate</u>:

$$\text{(IV.17)} \quad \sum_{k=0}^{m} \int_0^T \|\partial_t^k u\|^2_{m-k}\, dt \leq C\Big\{\|\vec{h}_\#\|^2_{m-\frac{1}{2}} + \int_0^T \|f(t)\|_0^2\, dt + \sum_{\ell=0}^{m-1} \int_0^T \|\partial_t^\ell u\|^2_{s'}\, dt\Big\} .$$

From this we can easily derive the standard versions of the coercive estimates. For instance, if we replace u by $\Lambda^s u$ we easily derive the estimate of $\int_0^T \|\partial_t^k u\|^2_{s+m-k}\, dt$ (which we could also have derived directly by substituting s + m for s + k in (IV.7)). We could also derive the "local" coercive estimates, by taking $T > 0$ small enough and using the standard inequality

$$\text{(IV.18)} \qquad \int_0^T \|\partial_t^\ell u\|^2_{s'}\, dt \leq T \int_0^T \|\partial_t^{\ell+1} u\|^2_{s'}\, dt \quad .$$

The local estimate is the valid for all $u \in C_c^\infty(X \times [0,T[)$:

$$\text{(IV.19)} \qquad \sum_{k=0}^{m} \int_0^T \|\partial_t^k u\|^2_{m-k}\, dt \leq C \left\{ \|\vec{h}_\#\|^2_{m-\frac{1}{2}} + \int_0^T \|f\|_0^2\, dt \right\}.$$

One may also want to "unravel" the composite vector $\vec{h}_\#$. The most favorable circumstances in which one can do this is when the degree, as a polynomial in ∂_t, of every $B_j(x, D_x, \partial_t)$ does not exceed m - 1. Then (cf. (III.41)) the Q_j are identically zero, and $\vec{h}_\# = (u_1, \ldots, u_\nu)$. We might also want to allow the orders of the B_j, as pseudodifferential operators in (x,t), not to be necessarily zero; this is usually the case when they are <u>differential</u> operators. Let m_j be the order of B_j; then what has been denoted by B_j should now be denoted by $\Lambda^{-m_j} B_j$. Under these hypotheses Estimate (IV.17) can be rewritten

$$\text{(IV.20)} \qquad \sum_{k=0}^{m} \int_0^T \|\partial_t^k u\|^2_{m-k}\, dt \leq C \left\{ \sum_{j=1}^{\bar{m}} \|B_j u|_{t=0}\|^2_{m-m_j-\frac{1}{2}} + \int_0^T \|Pu\|_0^2\, dt + \sum_{\ell=0}^{m-1} \int_0^T \|\partial_t^\ell u\|^2_{s'}\, dt \right\}.$$

As before s' is a real number as close to $-\infty$ as one wishes (C depends on the choice of s').

One might also have estimates for derivatives of <u>negative order</u> with respect to t. Etc. etc. On the subject of coercive estimates and their use to refine the measurement of hypo-ellipticity we refer to [9], Ch. 2.

V. THE $\bar{\partial}$-NEUMANN PROBLEM IN SUBDOMAINS OF $\mathbb{C}^N$

As a first example of noncoercive boundary value problem we shall describe now the $\bar{\partial}$-Neumann problem in an open subset Ω of $\mathbb{C}^N$, which we assume to be bounded and have a smooth (i. e., C^∞) boundary, X. Our aim is to obtain some information about the "boundary operator" $\mathcal{B}$ (see (III.49)). This information will enable us to apply Th. III.1 and derive from recent results about pseudodifferential operators with double characteristics that, under a suitable hypothesis, the $\bar{\partial}$-Neumann problem in Ω is hypo-elliptic. However we shall not motivate the interest in this boundary value problem; we shall not indicate how its hypo-ellipticity can be used to derive various results about holomorphic functions of several complex variables. On this subject we refer to [8].

Let us describe the notation used in the present chapter. The real coordinates in $\mathbb{R}^{2N}$ will be denoted by $x_1,\dots,x_N,y_1,\dots,y_N$; we shall write $z_j = x_j + \sqrt{-1}\, y_j$ $(1\le j\le N)$, and $z = (z_1,\dots,z_N)$, $x = \operatorname{Re} z$, $y = \operatorname{Im} z$ (note that x is a point of $\mathbb{R}^N$, identified with the subspace $y_1 =\dots= y_N = 0$ of $\mathbb{R}^{2N}$). We shall denote by $\alpha = \sum_{|J|=q} \alpha_J\, d\bar{z}_J$ a form of type $(0,q)$, say with smooth coefficients: this means that the α_J are C^∞ functions in $\mathbb{R}^{2N}$ (or in some subset of $\mathbb{R}^{2N}$) with values in $\mathbb{C}$, J ranges over the set of ordered multi-indices of length q: $0 \le q \le N$; if $q = 0$, α is a complex function; if $q\ge 1$, $J = (j_1,\dots,j_q)$ with $1 \le j_1 < \dots < j_q \le N$. If $\mathcal{F}$ is a space of scalar functions and $\mathcal{M}$ some subset of $\mathbb{C}^N$ we denote by $\Lambda^{0,q}\mathcal{F}(\mathcal{M})$ the space of the forms α as above with coefficients $\alpha_J \in \mathcal{F}(\mathcal{M})$: e. g., $\mathcal{F} = C^\infty$, C^∞_c, L^p, C^0, etc.; we may even extend this to distributions if it makes sense (depending on the nature of $\mathcal{M}$; $\mathcal{M}$ can be Ω, its closure $\overline{\Omega}$, the whole space $\mathbb{C}^N$, etc.). We recall the definition of the operator $\bar{\partial} : \Lambda^{0,q}C^\infty(\mathcal{M}) \to \Lambda^{0,q+1}C^\infty(\mathcal{M})$ $(q < N)$:

$$\text{(V.1)}\qquad \underline{\text{if}}\ \alpha = \sum_{|J|=q} \alpha_J\, d\bar{z}_J\ ,\qquad \bar{\partial}\alpha = \sum_{j=1}^{N} \sum_{|J|=q} \frac{\partial \alpha_J}{\partial \bar{z}_j}\, d\bar{z}_j \wedge d\bar{z}_J\ .$$

We may use the canonical basis $dx_I \wedge dy_J$ (I, J are ordered multi-indices of various lengths) in the exterior algebra over the dual of $\mathbb{R}^{2N}$ to extend the hermitian product and norm of $L^2(\mathbb{R}^{2N})$ to differential forms on $\mathbb{R}^{2N}$: if $f = \sum_{I,J} f_{I,J}\, dx_I \wedge dy_J$, $g = \sum_{I,J} g_{I,J}\, dx_I \wedge dy_J$, their L^2 inner product is $(f,g)_0 = \sum_{I,J} \int f_{I,J}\, \bar{g}_{I,J}\, dxdy$ (we assume that the coefficients $f_{I,J}$, $g_{I,J}$ are square integrable). With this definition, if α is a form of type $(0,q)$ as above, and $\beta = \sum_{|J'|=q'} \beta_{J'}\, d\bar{z}_{J'}$ is a form of type $(0,q')$, both with square-integrable coefficients, we see that $(\alpha,\beta)_0 = 0$ if $q \neq q'$, whereas, if $q = q'$,

$$\text{(V.2)}\qquad (\alpha,\beta)_0 = 2^q \sum_{|J|=q} \int \alpha_J\, \bar{\beta}_J\, dxdy$$

(in all this $dxdy = dx_1 \ldots dx_N dy_1 \ldots dy_N$ is the Lebesgue measure in $\mathbb{R}^{2N}$). We may then define the formal adjoint of $\bar{\partial}$ for the L^2 inner product (V.2). It is a linear operator $\vartheta : \Lambda^{0,q}C^\infty(\mathcal{M}) \to \Lambda^{0,q-1}C^\infty(\mathcal{M})$ $(q > 0)$:

$$\text{(V.3)}\qquad \vartheta\alpha = -2 \sum_{|J|=q-1} \sum_{\substack{k=1,\ldots,N \\ k \notin J'}} \varepsilon_J^{kJ'} \frac{\partial \alpha_J}{\partial z_k}\, d\bar{z}_{J'}\ ,$$

where $\alpha = \sum_{|J|=p} \alpha_J\, d\bar{z}_J$. In the summation at the right in (V.3) J is obtained by adjunction of $k \notin J'$ to J' and reordering; $\varepsilon_J^{kJ'} = +1$ or -1 according to the parity of the permutation that brings the set $\{k\} \cup J'$ to its ordered form, J (this notation will systematically be used from now on). The differential operator $\Box = \bar{\partial}\vartheta + \vartheta\bar{\partial}$ is called the <u>complex Laplacean</u> in $\mathbb{C}^N$; an easy computation shows that (with α as above)

$$\text{(V.4)} \qquad \Box\alpha = -\frac{1}{2} \sum_{|J|=q} \Delta\alpha_J \, d\bar{z}_J \, ,$$

where Δ is the ordinary Laplace operator in $\mathbb{R}^{2N}$, $\Delta = 4 \sum_{j=1}^{N} \frac{\partial^2}{\partial z_j \partial \bar{z}_j}$.

We select a C^∞ real valued function r in $\mathbb{C}^N$ such that X is defined by $r = 0$, with $dr \neq 0$ in a tubular neighborhood of X, and $r < 0$ in Ω.

The principal symbol $\sigma(\vartheta)(z, \zeta)$ of ϑ is a smooth function in the cotangent bundle over $\mathbb{R}^{2N}$ (which may be identified with $\mathbb{R}^{2N} \times \mathbb{R}^{2N}$) with values in the space of linear mappings $\Lambda^{0,q}\mathbb{C}^N \to \Lambda^{0,q-1}\mathbb{C}^N$ ($\Lambda^{0,q}\mathbb{C}^N$ is the obvious subspace of the complex exterior algebra over $\mathbb{R}^{2N}$). Therefore, if we substitute for ζ a C^∞ section of the cotangent bundle, such as $dr(z)$, we obtain a C^∞ section of the bundle of homomorphisms of $\Lambda^{0,q}T^{*\mathbb{C}}(\mathbb{R}^{2N})$ into $\Lambda^{0,q-1}T^{*\mathbb{C}}(\mathbb{R}^{2N})$ ($T^{*\mathbb{C}}$ stands for the complexified cotangent bundle). Since all bundles dealt with here are trivial we may identify $\sigma(\vartheta)(z,dr)$ with a matrix $(\sigma^J_{J'})_{|J|=q,\, |J'|\, q-1}$, which, by (V.3), is seen to be the following one:

$$\text{(V.5)} \qquad \sigma^J_{J'} = -2i\,\varepsilon^{kJ}_{J'} \frac{\partial r}{\partial z_k} \text{ if } J' = J \setminus \{k\}, \quad \sigma^J_{J'} = 0 \text{ if } J' \not\subset J .$$

Definition V.1.- The $\bar{\partial}$-Neumann problem in Ω is the problem of finding, given a form $f \in \Lambda^{0,q}C^\infty(\Omega)$, another form $u \in \Lambda^{0,q}C^2(\bar{\Omega})$ verifying

$$\text{(V.6)} \qquad \Box u = f \quad \text{in } \Omega,$$

$$\text{(V.7)} \qquad \sigma(\vartheta)(z,dr)u = 0 \quad \text{on } \partial\Omega,$$

$$\text{(V.8)} \qquad \sigma(\vartheta)(z,dr)\bar{\partial}u = 0 \text{ on } \partial\Omega .$$

We could have varied the regularity requirements on f, and for instance look at the case where $f \in \Lambda^{0,q}L^2(\Omega)$. But we shall assume, in fact, that f is C^∞ in the closure $\bar{\Omega}$ of Ω. The requirement that u is C^2 in $\bar{\Omega}$ insures that (V.7) & (V.8) will have a reasonable meaning. We shall actually seek to show that u is C^∞ up to pieces of the boundary $\partial\Omega$ where this is so for f (we shall have to make some hypotheses in order to obtain this).

By virtue of (V.5) we may rewrite (V.7) as follows (when $q > 0$):

(V.9) *When* $r = 0$ *we have, for all multi-indices* J', $|J'| = q - 1$,

$$\sum_{j \notin J'} \varepsilon_J^{jJ'} \frac{\partial r}{\partial z_j} u_J = 0 ,$$

where J *is the multi-index obtained by adjoining* j *to* J' *and reordering* (u_J is the corresponding coefficient of the form u).

As for (V.8) it can be rewritten as follows, provided q is > 0,

(V.10) *When* $r = 0$, *for all multi-indices* J, $J = q$, *we have*

$$\sum_{k \notin J, j \in J^*} \varepsilon_{J*}^{kJ} \varepsilon_{J*}^{jK} \frac{\partial r}{\partial z_k} \frac{\partial u_K}{\partial \bar{z}_j} = 0 ,$$

where J^* *is derived from* J *by adjoining* k *and reordering, while* $K = J^* \setminus \{j\}$.

When $q = 0$ (in which case u is a scalar function), (V.8) reads

(V.11) $$\bar{L}u = 0 \quad \text{when } r = 0,$$

with

(V.12) $$L = \sum_{k=1}^{N} \frac{\partial r}{\partial \bar{z}_k} \frac{\partial}{\partial z_k} .$$

Remark V.1.- If $q = N$, Condition (V.8) is of course void. Furthermore, to every J' of length $N - 1$ there is a unique $j \notin J'$, and thus (V.9) reads

(V.13) $$\frac{\partial r}{\partial z_j} u_{(1,\dots,N)} = 0 \quad \text{when } r = 0, \quad j = 1,\dots, N.$$

Since $dr \neq 0$ when $r = 0$, this is equivalent with

(V.13) $$u_{(1,\dots,N)} = 0 \quad \text{on } \partial\Omega .$$

Since Eq. (V.6) according to (V.4) reads

(V.14) $$\Delta u_{(1,\dots,N)} = -2f \quad \underline{\text{in}}\ \Omega ,$$

we see that, <u>when</u> $q = N$, <u>the</u> $\bar\partial$<u>-Neumann problem reduces to the Dirichlet problem</u>.

At this stage we localize the analysis near an arbitrary point of the boundary of Ω, which we shall take to be the origin in $\mathbb{C}^N$, and choose the coordinates in $\mathbb{C}^N$ so as to have, near the origin,

(V.15) $$r(z) = \operatorname{Re} z_N + 2 \operatorname{Re} \sum_{j,k=1}^{N-1} \frac{\partial^2 r(0)}{\partial z_j \partial z_k} z_j z_k + 2\mathcal{L}_0(r;z) + O(|z|^3),$$

where

(V.16) $$\mathcal{L}_0(r;w) = \sum_{j,k=1}^{N-1} \frac{\partial^2 r(0)}{\partial z_j \partial \bar z_k} w_j \bar w_k \qquad (w \in \mathbb{C}^N)$$

is called the <u>Levi form</u> of $\partial\Omega$ (or of Ω) at the point 0.

By availing ourselves of the expression (V.15) we are going to rewrite in a more practically useful fashion the boundary conditions (V.9), (V.10). We shall skip all computations, which are simple manipulations of indices. We regard the form $u = \sum_{|J|=q} u_J \, d\bar z_J$ as a vector, with components u_J. Actually we replace it by one of its linear transformations:

(V.17) $$u^{\#} = u - S(z)u,$$

defined as follows: When $q = 0$, $S(z) \equiv 0$. When $q > 0$,

(V.18) $$\underline{\text{If}}\ N \in J,\quad u_J^{\#} = u_J - (-1)^q \left(\frac{\partial r}{\partial z_N}\right)^{-1} \sum_{\substack{k=1\\ k\notin J}}^{N-1} \varepsilon_J^{kJ'} \frac{\partial r}{\partial z_k} u_K ,$$

<u>where</u> $J' = J \setminus \{N\}$ <u>and</u> K <u>is obtained by adjoining</u> k <u>to</u> J' ;

(V.19) $$\underline{\text{If}}\ N \notin J,\quad u_J^{\#} = u_J - 4|dr|^{-2} \sum_{j\in K,\, k\notin J'} \varepsilon_J^{jJ'} \varepsilon_K^{kJ'} \frac{\partial r}{\partial \bar z_j} \frac{\partial r}{\partial z_k} u_K ,$$

<u>with</u> $J' = J \setminus \{j\} = K \setminus \{k\}$.

In fact let us use the direct sum decomposition

(V.20) $$u^{\#} = u' + u'' ,$$

where $u'_J = 0$ if $N \in J$ and $u''_J = 0$ if $N \notin J$.

Proposition V.1.- Condition (V.9)(which presumes $q > 0$) is equivalent to:

(V.21) $$u'' = 0 \quad \text{when } r = 0.$$

In our reformulation of (V.10) we shall use the differential operator L defined by (V.12). We shall also use the matrix γ, acting on vectors of the kind u' (and transforming them into like vectors), defined by

(V.22) $$(\gamma u')_J = \sum_{j \in J, k \notin J'} \varepsilon_J^{jJ'} \varepsilon_K^{kJ'} \left(\frac{\partial^2 r}{\partial \bar{z}_j \partial z_k} - \frac{\partial^2 r}{\partial \bar{z}_j \partial z_N} \frac{\partial r}{\partial z_k} \Big/ \frac{\partial r}{\partial z_N} \right) u'_K ,$$

where $N \notin J'$, $J' = J \setminus \{j\} = K \setminus \{k\}$.

Noting that $u^{\#} = u'$ when $q = 0$ (there is no u" then) we state:

Proposition V.2.- Condition (V.10) (which presumes $q > 0$) is equivalent to

(V.23) $$(\bar{L} + \gamma) u' = 0 \quad \text{when } r = 0.$$

We may also take (V.23) as expressing the boundary condition (V.11) (which presumes $q = 0$) if we agree that $\gamma = 0$ when $q = 0$.

One may say that the $\bar{\partial}$-Neumann boundary conditions (V.7)-(V.8) are equivalent to the conjunction of Dirichlet conditions on the component u" of $u^{\#}$ with "Neumann-like" boundary conditions on the component u' .

Lastly we wish to rewrite Eq. (V.6) in terms of $u^{\#}$. We look closely at (V.18)-(V.19) and take into account the fact, following from (V.15), that

(V.24) $$\frac{\partial r}{\partial z_j} = O(|z|) \text{ if } j < N, \quad 2 \frac{\partial r}{\partial z_N} = 1 + O(|z|^2) , \quad z \sim 0 .$$

(We have already used above the fact that $\frac{\partial r}{\partial z_N} \neq 0$ in a neighborhood of 0.)

According to this we see that $S(0) = 0$ and that, consequently, $I - S(z)$ is invertible in a suitable neighborhood of the origin. We then set $u = (I - S(z))^{-1}u^{\#}$ in (V.6). This equation becomes

$$\Delta u^{\#} + Tu^{\#} = f^{\#}, \tag{V.25}$$

where

$$T = -2\,(I - S(z))\left[\Delta, (I - S(z))^{-1}\right]. \tag{V.26}$$

$$f^{\#} = -2\,(I - S(z))f. \tag{V.27}$$

It is now convenient to switch coordinates, from $\mathrm{Re}\, z_j$, $\mathrm{Im}\, z_j$ ($j = 1, \ldots, N$) to

$$x_j = \mathrm{Re}\, z_j,\ y_j = \mathrm{Im}\, z_j\ (j = 1,\ldots, N-1),\ r,\ y_N = \mathrm{Im}\, z_N. \tag{V.28}$$

Note that in these coordinates (near the origin) the boundary $X = \partial\Omega$ can be identified to a piece of the hyperplane of the coordinates x_j, y_k ($j < N$, $k \leq N$). Let us set

$$2\frac{\partial r}{\partial \bar{z}_j} = p_j + iq_j\ (p_j,\ q_j \text{ real; } j = 1,\ldots, N)\ ;\ R = |dr|;\ h = \Delta r. \tag{V.29}$$

By virtue of (V.24) we have:

$$p_j = O(|z|),\ q_j = O(|z|) \text{ if } j < N;\quad p_N = 1 + O(|z|^2), \tag{V.30}$$
$$q_N = O(|z|^2),\quad R = 1 + O(|z|^2).$$

Let us introduce the additional notation:

$$M_0 = \sum_{j=1}^{N-1} \left(p_j \frac{\partial}{\partial x_j} + q_j \frac{\partial}{\partial y_j}\right) + q_N \frac{\partial}{\partial y_N}, \tag{V.31}$$

$$M_1 = \sum_{j=1}^{N-1} \left(q_j \frac{\partial}{\partial x_j} - p_j \frac{\partial}{\partial y_j}\right) - p_N \frac{\partial}{\partial y_N}. \tag{V.32}$$

An easy computation shows that

$$\Delta = R^2 \frac{\partial^2}{\partial r^2} + 2M_0 \frac{\partial}{\partial r} + h \frac{\partial}{\partial r} + \Delta',\quad 4L = R^2 \frac{\partial}{\partial r} + M_0 + iM_1, \tag{V.32}$$

where Δ' is the Laplace operator on X: $\Delta' = \sum_{j=1}^{N-1} \left(\frac{\partial^2}{\partial x_j^2} + \frac{\partial^2}{\partial y_j^2}\right) + \frac{\partial^2}{\partial y_N^2}$.

From (V.26) and (V.32) we see that

(V.33) $$T = h_1 \frac{\partial}{\partial r} + T' ,$$

where h_1 is a C^∞ matrix-valued function near 0 and T' a first-order differential operator, with C^∞ matrix-valued coefficients, whose principal part is tangential to X. We factorize directly Eq. (V.25):

(V.34) $$I \quad + T = (IR \frac{\partial}{\partial r} + A_1)(IR \frac{\partial}{\partial r} - A).$$

<u>Here as in the sequel we completely disregard the error terms coming from regularizing operators</u>. They can be handled exactly as in Ch. III, Ch. IV. And they have no effect on the reasonings nor on their conclusions. We get:

(V.35) $$A_1 = RAR^{-1} + (2R^{-1}M_0 + R^{-1}h - \frac{\partial R}{\partial r})I + R^{-1}h_1 ,$$

(V.36) $$A^2 + \frac{2}{R}M_0 A + (\frac{h}{R} + \frac{h_1}{R} - \frac{\partial R}{\partial r} - [A,R]R^{-1})A + R\left[\frac{\partial}{\partial r},A\right] + I\Delta' + T' = 0.$$

Eq. (V.36) shows that A is a pseudodifferential operator of order one on X, depending smoothly on r (this implies $\left[\frac{\partial}{\partial r},A\right] = \frac{\partial A}{\partial r}$). We may and shall take the principal symbol of A to be

(V.37) $$\sigma(A) = - \sigma(M_0)/R + \{ \sigma(M_0)^2/R^2 - \sigma(\Delta')\}^{1/2} .$$

It will be shown below that Re $\sigma(A) > 0$ in the complement of the zero section in T^*X. Since M_0 is a real vector field on X, its principal symbol $\sigma(M_0)$ is purely imaginary; we shall see that $\sigma(-\Delta') + \sigma(M_0)/R^2 > 0$. The square-root in (V.37) is the positive one. From (V.35) & (V.37) we get

(V.38) $$\sigma(A_1) = \sigma(M_0)/R + \{\sigma(M_0)^2/R^2 - \sigma(\Delta')\}^{1/2} .$$

We decompose the equation (V.25) into the system:

(V.39) $$R\frac{\partial u^\#}{\partial r} - Au^\# = v^\# ,$$

(V.40) $$R\frac{\partial v^\#}{\partial r} + A_1 v^\# = f^\# ,$$

to which we must adjoin the boundary conditions (V.21), (V.23), which we rewrite here:

(V.41) $$R\frac{\partial u'}{\partial r} + \frac{1}{R}(M_0 - iM_1 + 4\gamma)u' = 0\,, \qquad u'' = 0 \text{ \underline{when} } r = 0.$$

Let us then define the pseudodifferential operator on X, A_0 , with values in the space of matrices that transform vectors of the kind u' into like ones, as follows:

(V.42) $$A_0 u'(0) = (Au^{\#})'\big|_{r=0} \quad .$$

If we extract $R\frac{\partial u'}{\partial r}$ from (V.39) and put it into (V.41), we may rewrite the latter as

(V.43) $$\mathcal{B}'u'(0) = - v'(0)\,, \quad u''(0) = 0\,,$$

where u'(0), u"(0), v'(0) are the values of u', u", v' at r = 0 (these values are functions in X) and where

(V.44) $$\mathcal{B}' = A_0 + R^{-1}\left[I(M_0 - iM_1) + 4\gamma\right]\big|_{r=0} \quad .$$

The local representation of the $\bar{\partial}$-Neumann problem (V.21)-(V.23)-(V.25) provided by (V.39)-(V.40)-(V.43) is the analogue of the decomposition of (*) into (**)-(***) in Ch. III. It should indeed be noted that the equations (V.39)-(V.40) must be satisfied in the portion $r < 0$ of a neighborhood of the origin, and that the role of the variable t in Ch. III will here be played by - r.

<u>Remark V.2</u>.- When q = 0, all the matrices S(z), γ , T, T', h_1 vanish identically, and the pseudodifferential operators A, A_0 , $\mathcal{B}'$ are scalar - as they should be.

The principal symbol of the boundary operator $\mathcal{B}'$

We shall use the notation $\zeta_j = \xi_j + i\eta_j$ (j = 1,..., N-1), $\zeta_N = i\eta_N$. Then we have

(V.45) $$\sigma(-\Delta') = |\zeta|^2\,, \quad \sigma(M_0 - iM_1) = 2i\,\zeta\cdot\partial r\,,$$

We derive at once from (V.37), (V.44) and (V.45):

(V.46) $$\sigma(\mathcal{B}') = \{|\zeta|^2 - (\mathrm{Re}\,\zeta\cdot\partial r/|\partial r|)^2\}^{1/2} - \mathrm{Im}\,\zeta\cdot\partial r/|\partial r| .$$

In all this ∂r is the "vector" with components $\partial r/\partial z_j$ ($j = 1,\ldots, N$); we observe that $R|_{r=0} = 2\,|\partial r|\,|_{r=0}$. By virtue of (V.46) we see that $\sigma(\mathcal{B}') \geq 0$. Let us multiply $\sigma(\mathcal{B}')$ by

(V.47) $$B_0 = \{|\zeta|^2 - (\mathrm{Re}\,\zeta\cdot\partial r/|\partial r|)^2\}^{1/2} + \mathrm{Im}\,\zeta\cdot\partial r/|\partial r| ,$$

and set

(V.48) $$F = |\partial r|^2 B_0\,\sigma(\mathcal{B}') = |\zeta|^2|\partial r|^2 - |\zeta\cdot\partial r|^2 .$$

The <u>characteristic</u> <u>set</u> of $\mathcal{B}'$, <u>i</u>. <u>e</u>., the zero-set of $\sigma(\mathcal{B}')$, is contained in the set

(V.49) $$|\zeta\cdot\partial r| = |\zeta|\,|\partial r| \qquad \mathrm{Im}\,\zeta\cdot\partial r \geqslant 0 .$$

The first-one of these of these conditions requires $\zeta = c\,\overline{\partial r}$ for some complex function c . By the fact that $\zeta_N = i\eta_N$ this in turn requires $c = i\eta_N/\dfrac{\partial r}{\partial \bar z_N}$ (cf. (V.24)), and therefore (V.49) implies

(V.50) $$\xi_j + i\eta_j - i\eta_N \frac{\partial r}{\partial \bar z_j}\Big/\frac{\partial r}{\partial \bar z_N} = 0 , \ j = 1,\ldots, N-1 .$$

We also observe that the second condition (V.49) can be rewritten $\mathrm{Im}\,c > 0$ which, for z small, is equivalent with $\eta_N \geqslant 0$. But if $\eta_N = 0$, (V.50) implies $\xi_j = \eta_j = 0$ for all $j = 1,\ldots, N-1$, hence if we restrict the concept of characteristic set to the complement of the zero section in the cotangent bundle (here over a portion of X) we see that we must complement (V.50) with

(V.51) $$\eta_N > 0 .$$

<u>In the region</u> $\eta_N < 0$ $\mathcal{B}'$ <u>is elliptic</u>. Note also that, in the region (V.51), $B_0 > 0$. Recalling that $\dim T^*X = 2(2N - 1)$ we state:

Proposition V.3.- Char $\mathcal{B}'$ is a C^∞ submanifold of dimension $2N$ of $T^*X \setminus 0$. If z_0 is sufficiently close to the origin (in X), the intersection of Char $\mathcal{B}'$ with the fibre $T^*_{z_0}X$ is a single ray $\xi_j = \rho\xi_j^0$, $\eta_j = \rho\eta_j^0$ $(j = 1,\dots, N-1)$, $\eta_N = \rho > 0$.

When $N = 1$ there are no equations (V.50); Char $\mathcal{B}'$ is defined simply by the inequality (V.51). In the remainder of this section we suppose $N \geq 2$.

Let us introduce the following complex vector fields:

$$Z_j = \frac{\partial}{\partial z_j} - \left(\frac{\partial r}{\partial z_j} \Big/ \frac{\partial r}{\partial z_N}\right) \frac{\partial}{\partial z_N} \quad , \; j = 1,\dots, N-1 . \tag{V.52}$$

The complex conjugates $\bar{Z}_1, \dots, \bar{Z}_{N-1}$ define what is called the induced Cauchy-Riemann operator on X ; for more information on this important topic we refer to Ch. V of [8].

It is seen at once that Eqq. (V.50) can be rewritten:

$$\sigma(Z_j) = 0 \quad , \quad j = 1,\dots, N-1 . \tag{V.53}$$

On the other hand let us denote by w the $(N-1)$-vector with components $w_j = \frac{\partial r}{\partial z_j} \Big/ \frac{\partial r}{\partial z_N}$, and by $\sigma(Z)$ the one with components $\sigma(Z_j)$. Then

$$(2\,|\partial r|)^{-2} F = |\sigma(Z)|^2 - |w.\overline{\sigma(Z)}|^2/(1 + |w|^2) . \tag{V.54}$$

Thus, in the region $\eta_N > 0$, we have (cf. (V.47)):

$$\sigma(\mathcal{B}') = 4B_0^{-1} \sum_{j=1}^{N-1} |\sigma(Z_j)|^2 - \sum_{j,k=1}^{N-1} c_{jk}\,\sigma(Z_j)\,\overline{\sigma(Z_k)} , \tag{V.55}$$

where (c_{jk}) is a self-adjoint positive semidefinite $(N-1)\times(N-1)$ matrix dependending smoothly on the variable point in a suitable neighborhood of the origin, in the base X. Since, by (V.24), $w_j(0) = 0$ for all $j < N$, we have:

$$c_{jk}(0) = 0 , \quad \forall\; j, k = 1,\dots, N-1 . \tag{V.56}$$

We derive from all this:

Proposition V.4.- The principal symbol $\sigma(\mathcal{B}')$ vanishes exactly of order two on the (smooth) manifold Char $\mathcal{B}'$.

Remark V.3.- Let us show that $\sigma(A)$ is >0 on the portion of $T^*X \setminus 0$ which lies over a sufficiently small neighborhood of the origin in X (and for small enough values of $-r$). We derive from (V.37) (using the notation of (V.45)):

(V.57) $$\sigma(A) = \{|\zeta|^2 - (\mathrm{Re}\,\zeta\cdot\partial r/|\partial r|)^2\}^{1/2} + i\,\mathrm{Re}\,\zeta\cdot\partial r/|\partial r| .$$

Thus Re $\sigma(A) \geq 0$ near 0 and Re $\sigma(A) = 0$ only if Re $\zeta\cdot\partial r = |\zeta||\partial r|$ and therefore Im $\zeta\cdot\partial r = 0$. But we have just seen that this conjunction implies $\zeta = 0$. Q. E. D.

The subprincipal symbol of the "boundary operator" $\mathcal{B}'$

We continue to represent the variable point in the cotangent bundle over X by $(x_1,\ldots,x_{N-1},y_1,\ldots,y_N,\xi_1,\ldots,\xi_{N-1},\eta_1,\ldots,\eta_N)$. Since we are dealing here with classical pseudodifferential operators we can consider their total symbol which is a formal series of symbols that are positive homogeneous of integral degreees with respect to the fibre variable (ξ,η). Thus the total symbol of A is $a_0 + a_1 + \ldots$, with deg $a_j = 1 - j$, and the one of $\mathcal{B}'$ is $b_0 + b_1 + \ldots$ with deg $b_j = 1 - j$ also $(j = 0, 1,\ldots)$. By definition, the subprincipal symbol of $\mathcal{B}'$ is the quantity:

(V.58) $$\sigma_1(\mathcal{B}') = b_1 - \frac{1}{2i}\left\{\sum_{j=1}^{N-1} \frac{\partial^2 b_0}{\partial x_j\partial\xi_j} + \sum_{k=1}^{N} \frac{\partial^2 b_0}{\partial y_k\partial\eta_k}\right\} .$$

The subprincipal symbol is easily shown to be invariant under coordinates change - provided it is restricted to the set of zeros where the principal symbol and all its first derivatives vanish. By Prop. V.4 the latter is exactly what happens in the case of $\mathcal{B}'$, and therefore the restriction of $\sigma_1(\mathcal{B}')$ to Char $\mathcal{B}'$ is invariant.

The computation of $\sigma_1(\mathcal{B}')$ is routine: it is based on the information provided by (V.30), (V.36) and, of course, the definition of $\mathcal{B}'$, (V.44). Let us here content ourselves with giving its value at a point $\omega_o \in$ Char $\mathcal{B}'$ which lies directly above the origin (i. e., in the cotangent space to X at 0). By Prop. V.3 such a point is completely determined by requiring that $\eta_N = 1$ (since $\sigma_1(\mathcal{B}')$ is positive-homogeneous of degree zero, it does not really matter what value of $\eta_N > 0$ we choose). Recalling that $h = \Delta r$, one easily finds

$$\sigma_1(\mathcal{B}') = 4\gamma(0) - \tfrac{1}{2}\Delta r(0)I \quad \underline{\text{at}} \quad \omega_o . \tag{V.59}$$

We recall that γ is the matrix defined in (V.22). Its generic entry is a scalar γ_J^K where J, K are multi-indices of length q such that $N \notin J$, $N \notin K$. We can compute γ_J^K by using (V.22). But the computation of this, and of $\sigma_1(\mathcal{B}')$ is made easier if we assume that the Levi form (V.16) has been diagonalized at the origin:

$$\frac{\partial^2 r(0)}{\partial z_j \partial \bar{z}_k} = \lambda_j \delta_{jk} \quad (j, k = 1, \dots, N-1) , \tag{V.60}$$

which is always possible by a linear change of coordinates in $\mathbb{C}^N$. Then

$$\gamma_J^K(0) = \sum_{j \in J} \lambda_j \quad \underline{\text{if}}\ J = K, \qquad \gamma_J^K(0) = 0 \ \underline{\text{if}}\ J \neq K. \tag{V.61}$$

Also $-\frac{1}{2}\Delta r(0) = -2 \sum_{j=1,\dots,N-1} \lambda_j$.

<u>Proposition V.5</u>.- <u>Suppose that</u> (V.60) <u>holds. The restriction of</u> $\sigma_1(\mathcal{B}')$ <u>to the intersection of</u> Char $\mathcal{B}'$ <u>with the cotangent space to</u> X <u>at the origin is a diagonal matrix with diagonal entries equal to</u>

(V.62) $$2\left(\sum_{j\in J}\lambda_j - \sum_{j\notin J}\lambda_j\right)$$

<u>where</u> J <u>ranges over the collection of multi-indices with length</u> q <u>which do not contain</u> N.

<u>Hypo-ellipticity with loss of one derivative. Condition Z(q)</u>

Prop. V.3, V.4, V.5 enable us to use recent results of various authors (mainly see [2], [7]) to obtain necessary and sufficient conditions for $\mathcal{B}'$ to be <u>hypo-elliptic with loss of one derivative</u>, which means that for any open subset $\mathcal{O}$ of the boundary X, any distribution u' in $\mathcal{O}$ and any real number s,

(V.63) $$\mathcal{B}'u' \in H^s_{loc}(\mathcal{O};V') \quad \underline{\text{implies}} \quad u' \in H^s_{loc}(\mathcal{O};V')$$

(we have denoted by V' the space of vectors of the kind u'). Property (V.63) indeed evidences the loss of one order of smoothness, since $\mathcal{B}'$ is of order <u>one</u>: if $\mathcal{B}'$ were elliptic, which is taken to be the case of <u>no</u> derivative loss, $\mathcal{B}'u'$ in H^s_{loc} would imply u' in H^{s+1}_{loc}. Property (V.63) is equivalent with local estimates (valid provided that the open set $\mathcal{O}$ is small enough):

(V.64) $$\|u'\|_s \le \text{const.}\,\|\mathcal{B}'u'\|_s \;, \qquad u' \in C^\infty_c(\mathcal{O};V') \,.$$

If (V.64) holds one can show that the so-called $\frac{1}{2}$-estimate holds for the $\bar{\partial}$-Neumann problem: all one has to do is to duplicate the argument in Ch. IV ("Coercive estimates"), using now the operator $\mathcal{B}(u' + u'') = \mathcal{B}'u'$ (and, as in Ch. IV. taking advantage of the estimates of Ch. II for the parametrix of the heat equation; here the role of the operator P of Ch. IV is played by Δ + T ; see (V.25)).

We continue to use the notation of the preceding section. In particular ω_o is the point in Char $\mathcal{B}'$ defined by $z = 0$, $\eta_N = 1$. We shall denote by E_o the <u>tangent</u> space to T^*X at ω_o, and by $E_o^{\mathbb{C}}$ its complexification.

The Taylor expansion of $\sigma(\mathcal{B}')$ about ω_o begins with the quadratic form

(V.65) $$\frac{1}{2}\sum_{j=1}^{N-1} |\sigma_j|^2$$

where, for each $j < N$, σ_j is the linear part of $\frac{2}{i}\overline{\sigma(Z_j)}$ at ω_o (cf.(V.52)):

(V.66) $$\sigma_j = \zeta_j - 2i\lambda_j z_j - \sum_{k=1}^{N-1} \frac{\partial^2 r(0)}{\partial \bar{z}_j \partial \bar{z}_k} \bar{z}_k .$$

We denote by $Q(\theta,\theta')$ the bilinear form on E_o defined by the quadratic form (V.65); we extend it bilinearly (not sesquilinearly!) to $E_o^{\mathbb{C}}$. We see that

(V.67) $$Q(\theta,\bar{\theta}) \geq 0 , \qquad \forall\, \theta \in E_o^{\mathbb{C}} .$$

We introduce now the canonical symplectic form on E_o,

(V.68) $$\Sigma = \sum_{j=1}^{N-1} d\xi_j \wedge dx_j + \sum_{j=1}^{N} d\eta_j \wedge dy_j ,$$

and also extend it as a bilinear form to $E_o^{\mathbb{C}}$. Since Σ is nondegenerate there is an endomorphism of $E_o^{\mathbb{C}}$, Φ, such that

(V.69) $$Q(\alpha,\beta) = i\, \Sigma(\alpha,\Phi\beta) , \quad \alpha,\beta \in E_o^{\mathbb{C}} .$$

The following is not difficult to prove:

<u>Proposition V.6.- The eigenvalues of the endomorphism Φ are the real numbers</u> $2\lambda_j$, $-2\lambda_j$ $(j = 1,\ldots, N-1)$ (where, we recall, the λ_j are the eigenvalues of the Levi form of X at the origin).

We come now to the results in [2], [7]. They tell us that if the open set $\mathcal{O}$ contains the origin and if (V.64) holds, then we must have the following:

(V.70) Let χ_j ($j = 1,\ldots,r$) be the positive eigenvalues of Φ. Then, whatever the eigenvalue μ of $\sigma_1(\mathcal{B}')$, the vector θ in $E_o^{\mathbb{C}}$ such that $\Phi^d\theta = 0$ for some $d > 0$, the r-tuple $m_1,\ldots, m_r$ of nonnegative integers,

$$\mu + Q(\theta,\bar{\theta}) + \sum_{j=1}^{r}(2m_j + 1)\chi_j \neq 0 .$$

According to Prop. V.6 we may take $\chi_j = 2|\lambda_j|$, $j = 1,\ldots, r$ (possibly after changing the indices of the λ_j); of course $r < N$, and $\lambda_j = 0$ if $r < j \leqslant N - 1$. From Prop. V.5 we derive:

(V.71) For all eigenvalues μ of $\sigma_1(\mathcal{B}')$, $\mu + \sum_{j=1}^{r}\chi_j \geqslant 0$.

If then we also take into account (V.67) we see that Condition (V.70) is equivalent to

(V.72) Whatever the eigenvalue μ of $\sigma_1(\mathcal{B}')$, $\mu + \sum_{j=1}^{r}\chi_j > 0$.

It is obvious that Condition (V.72), if it holds at ω_o, will also hold in a full neighborhood of ω_o (in Char $\mathcal{B}'$). Since (V.67) also holds in a neighborhood of ω_o, we derive that Condition (V.70) will hold in a full neighborhood of ω_o as soon as (V.72) holds at ω_o. The results in [2], [7] then tell us that the estimates (V.64) holds if the set $\mathcal{O}$ is small enough.

Now, according to Prop. V.5, (V.72) can be restated by saying that

(V.73) $$\sum_{j\in J}\lambda_j - \sum_{j\notin J}\lambda_j + \sum_{j=1}^{N-1}|\lambda_j| > 0$$

whatever the multi-index J, of length q, such that $N \notin J$.

Property (V.73) can, in turn, be rephrased as follows:

$Z(q)_0$ The Levi form of X at the origin has at least $N - q$ eigenvalues which are > 0 or at least $q + 1$ which are < 0.

Thus, according to the main result in [2], [7], we may state:

Theorem V.1.- In order that $\mathcal{B}'$ be hypo-elliptic with loss of one derivative in some open neighborhood of the origin in X it is necessary and sufficient that Condition $Z(q)_0$ hold.

As we have said the estimate (V.64) implies the $\frac{1}{2}$-estimate for the $\bar{\partial}$-Neumann problem. Another proof of the $\frac{1}{2}$-estimate can be found in [8], Ch. III.

Remark V.4.- When $q = N$ Condition $Z(q)$ is trivially satisfied. We know that in this case the $\bar{\partial}$-Neumann problem reduces to the Dirichlet problem.

Remark V.5.- The open set Ω (or its boundary X) is said to be strongly pseudoconvex at the point $z = 0$ if the Levi form at that point, (V.16), is positive-definite, i. e., every one of its eigenvalues λ_j is > 0. In this case it is clear that $Z(q)_0$ holds provided $q \geq 1$. For $q = 0$ it does not: the operator $\mathcal{B}'$ is not hypo-elliptic (with any regularity loss!) when $q = 0$ and Ω is strongly pseudoconvex.

VI. BOUNDARY VALUE PROBLEMS OF PRINCIPAL TYPE

Th. V.1 shows that, in the $\bar{\partial}$-Neumann problem, the hypo-ellipticity is determined by the subprincipal symbol of the boundary operator $\mathcal{B}$. There are problems in which it is dependent only on properties of the principal symbol of $\mathcal{B}$. This is the case when $\mathcal{B}$ is of principal type.

Let us describe what this means. We shall assume, as in Ch. IV, that $\nu = m^-$, i. e., the number of boundary conditions is equal to the number of roots of the polynomial in z, $\sigma(P)(x,t,\xi,z)$, with real part > 0.

Since we also assume that dim $H < +\infty$ we shall regard $\mathcal{B}$ as a pseudodifferential operator in X with values in the space of $r \times r$ matrices, for some integer $r \geq 1$ (these are matrices with scalar entries).

Definition VI.1.- We say that the boundary value problem (*) (Ch. III) is of principal type if $\nu = m^-$ and if the following holds:

$$\text{(VI.1)} \qquad \forall (x,\xi) \in T^*X \setminus 0 \text{ , } \det \sigma(\mathcal{B})(x,\xi) = 0 \text{ implies}$$

$$d_\xi \det \sigma(\mathcal{B})(x,\xi) \neq 0 \text{ .}$$

The characteristic set of $\mathcal{B}$, Char $\mathcal{B}$, is the subset of $T^*X \setminus 0$ where $\sigma(\mathcal{B})(x,\xi)$ is not invertible, i. e.,

$$\text{(VI.2)} \qquad \text{Char } \mathcal{B} = \{ (x,\xi) \in T^*X \setminus 0 \text{ ; } \det \sigma(\mathcal{B})(x,\xi) = 0 \}$$

When (VI.1) holds Char $\mathcal{B}$ is a subset of a smooth submanifold of $T^*X \setminus 0$ of codimension one, of course conic. When $\sigma(\mathcal{B})$ is real, it is such a submanifold, otherwise it is (in general) a proper subset of such a manifold.

Let $b^{¢}(x,\xi)$ denote the cofactor matrix of $\sigma(\mathcal{B})(x,\xi)$, $\mathcal{B}^{¢}$ any classical pseudodifferential operator in X with principal symbol $b^{¢}$, B any such operator with principal symbol I det $\sigma(\mathcal{B})$. It is clear that $\mathcal{B}^{¢}\mathcal{B}$ and $\mathcal{B}\mathcal{B}^{¢}$ differ from B by a (matrix-valued) operator of degree $\leq$ deg B - 1. Since the hypo-ellipticity of $\mathcal{B}^{¢}\mathcal{B}$ (resp., the solvability of $\mathcal{B}\mathcal{B}^{¢}$) implies that of $\mathcal{B}$, in many questions one may assume that the principal symbol of the pseudodifferential operator under study is a scalar multiple of the identity matrix. Thus the statements below concern the operator B and we write $\sigma(B) = b(x,\xi)I$. When applying these results to $\mathcal{B}$ we take $b = \det \sigma(\mathcal{B})$. Note that this argument, applied to most elliptic systems, shows that they are reducible to the type (III.1).

When $\mathcal{B}$ is of principal type (which the preceding argument does not presume) one can reduce, at least microlocally, the study of $\mathcal{B}$ to that of an <u>entirely</u> scalar operator, also of principal type. Indeed, the rank of the $r \times r$ matrix $\sigma(\mathcal{B})$ at any $(x,\xi) \in \mathrm{Char}\,\mathcal{B}$ is exactly $r-1$ and one can find an <u>elliptic</u> matrix-valued operator $\mathcal{M}$ such that, in a conic neighborhood of (x,ξ), $\mathcal{M} \circ \mathcal{B}$ is triangular, with r-1 diagonal entries of the form

(VI.3) $$I + S, \quad \text{with } S \in \Psi^{-1}(X),$$

the r-th one being of principal type. This remains true even when "principal type" is taken to mean

(VI.4) $$\forall (x,\xi) \in \mathrm{Char}\,\mathcal{B}, \qquad d_{x,\xi} \det \sigma(\mathcal{B})(x,\xi) \neq 0 .$$

Let $a(x,\xi)$ be a C^∞ complex function on $T^*X \setminus 0$. The <u>Hamiltonian field</u> of a is the (complex) vector field on $T^*X \setminus 0$,

(VI.5) $$H_a = \sum_{j=1}^{n} \frac{\partial a}{\partial \xi_j} \frac{\partial}{\partial x_j} - \frac{\partial a}{\partial x_j} \frac{\partial}{\partial \xi_j} .$$

By a <u>bicharacteristic strip</u> of a we mean a C^1 curve $]0,1[\ni t \mapsto \gamma(t) \in T^*X \setminus 0$ such that, for all $0<t<1$, $\frac{d\gamma}{dt}(t)$ is $\neq 0$ and proportional to the vector field H_a at the point $\gamma(t)$ (in particular, γ is a "true" curve). Since a is complex there will not be, in general, bicharacteristic strips of a (even if $da \neq 0$). But if a is real there is a unique bicharacteristic strip of a through every point $(x^o,\xi^o) \in T^*X \setminus 0$ such that $da(x^o,\xi^o) \neq 0$. Whenever such a strip exist we have $a = $ const. along it, since $H_a a = 0$.

<u>Definition VI.2.- Let $\mathcal{O}$ be an open subset of</u> $T^*X \setminus 0$. <u>We say that Condition</u> (Ψ') <u>is satisfied by</u> B <u>in</u> $\mathcal{O}$ <u>if there exists a</u> C^∞ <u>function</u> q <u>in</u> $\mathcal{O}$ <u>such that the following is true:</u>

(VI.6) $$\forall (x,\xi) \in \mathcal{O}, \qquad b(x,\xi) = 0 \implies d\ \mathrm{Re}(qb)(x,\xi) \neq 0 ;$$

(VI.7) <u>along any bicharacteristic strip of</u> Re(qb) <u>contained in</u> $\mathcal{O}$ <u>on</u>

which $\mathrm{Re}(qb) = 0$, if $\mathrm{Im}(qb) > 0$ at some point, $\mathrm{Im}(qb) \geq 0$ at any later point (for the orientation of $H_{\mathrm{Re}(qb)}$).

We say that B satisfies the strict condition (Ψ') in $\mathcal{O}$ if the function q can be selected so as to satisfy, in addition to (VI.6) & (VI.7), the following:

(VI.8) b does not vanish identically on any bicharacteristic strip of Re (qb) contained in $\mathcal{O}$ on which $\mathrm{Re}(qb) = 0$.

We say that B satsifies the condition (Ψ') (resp., the strict condition (Ψ')) at a point (x^o, ξ^o) of $T^*X \setminus 0$ if it does in some open neighborhood of (x^o, ξ^o).

It can be shown (see [11], [15]) that if (VI.6) & (VI.7) (resp., & (VI.8)) hold for some function q, then (VI.7) (resp., & (VI.8)) will hold fo any other function which satisfies (VI.6). (About (VI.8) see [14] .)

It is reasonable to conjecture that

(VI.9) if the operator B satisfies the following condition:

(VI.10) $\forall\ (x, \xi) \in T^*X \setminus 0\ ,\ b(x, \xi) = 0 \Longrightarrow db(x, \xi) \neq 0\ ,$

then, in order that B be hypo-elliptic in X, it is necessary and sufficient that B satisfy the strict condition (Ψ') at every point of $T^*X \setminus 0$.

Observe that Property (VI.10) is weaker than the property of being of principal type, which reads

(VI.10) $\forall\ (x, \xi) \in \mathrm{Char}\ B,\quad d_\xi\, b(x, \xi) \neq 0\ .$

(For db stands for the differential of b with respect to (x, ξ).) Let us indicate briefly to what extent the conjecture (VI.9) has been proved.

Differential operators

In this section we suppose that B is a differential operator; this is difficult to reconcile with the fact that B is of order zero, but what we mean is that B becomes a differential operator after multiplication by a suitable elliptic operator (of the kind Λ^m for some integer $m \geqslant 0$). We assume the multiplication effected. Since then $b(x,-\xi) = (-1)^m b(x,\xi)$ if we assume, as we shall, that the open set $\mathcal{O}$ in Def. VI.2 contains every fibre T_x^*X which it intersects, Condition (VI.7) must be replaced by

(VI.11) along any bicharacteristic strip of Re(qb) contained in on which Re(qb) = 0, Im(qb) does not change sign.

Definition VI.3.- We say that B (assumed now to be a classical pseudo-differential operator in X, not necessarily a differential operator) satisfies Condition (P) in the open subset $\mathcal{O}$ of $T^*X \setminus 0$ if there is a C^∞ function q in $\mathcal{O}$ such that (VI.6) & (VI.11) hold.

Now it is clear that if both (VI.6)-(VI.11) and (VI.8) hold we have:

(VI.12) Im(qb) keeps the same sign on each connected component of the zero-set of Re(qb) in $\mathcal{O}$.

Theorem VI.1.- Let B be a differential operator in X. Suppose that each point in $T^*X \setminus 0$ has an open neighborhood $\mathcal{O}$ in which there is a smooth function q such that $d_\xi \operatorname{Re}(qb) \neq 0$ on Char B, and such that furthermore (VI.8) and (VI.12) hold. Then B is hypo-elliptic in X.

Suppose that B is of principal type (i. e., (VI.10) holds) and that there is an open set $\mathcal{O} \subset T^*X \setminus 0$ and a C^∞ function q in $\mathcal{O}$ such that either (VI.8) is not true or there is a bicharacteristic strip of Re(qb) in $\mathcal{O}$, on which Re(qb) = 0 and at some point of which Im(qb) *has a zero of odd order* (along the bicharacteristic in question). *Then B is not hypo-elliptic in* X.

Th. VI.1 is proved in [14].

Subelliptic operators

This is an interesting subclass of hypo-elliptic operators.

Definition VI.4.- A pseudodifferential operator A *of order* m *in* X *is said to be subelliptic if every point* x^o *in* X *has an open neighborhood* Ω *such that, for some number* $\delta = \delta(x^o) > 0$, *the following is true:*

(VI.13) $$\|u\|_{m-1+\delta} \leq \text{const.} \|Au\|_0 , \qquad \forall\, u \in C_c^\infty(\Omega) .$$

(We suppose made some choice of the Sobolev norms in X.) It is easy to see that (VI.13) implies that, whatever s real, there is an open neighborhood Ω_s of x^o and a constant $C_s > 0$ such that

(VI.14) $$\|u\|_{s+m-1+\delta} \leq C_s \|Au\|_s , \qquad \forall\, u \in C^\infty(\Omega_s) ,$$

or that, whatever the compact set $K \subset X$, the real numbers s, s', there is a constant $C = C(s,s',K) > 0$ such that

(VI.15) $$\|u\|_{s+m-1+\delta} \leq C_s(\|Au\|_s + \|u\|_{s'}), \quad \forall\, u \in C^\infty(K) .$$

It is also easy to check that if A is subelliptic in X, we have

(VI.16) $$WF(Au) = WF(u) , \qquad \forall\, u \in \mathcal{D}'(X),$$

hence A is, in particular, hypo-elliptic.

In applying this notion of subellipticity to our operator B we should consider distributions with values in $\mathbb{C}^r$; the extension is routine.

Definition VI.5.- Let k be an integer ≥ 0. We say that B satisfies the condition (Ψ'_k) in the open set $\mathcal{O} \subset T^*X \setminus 0$ there is $q \in C^\infty(\mathcal{O})$ such that (VI.6) & (VI.7) hold, and also the following:

(VI.17) along every bicharacteristic strip of Re(qb) contained in $\mathcal{O}$, on which Re(qb) = 0, Im(qb) has zero of order at most k.

Note that (Ψ'_k) implies the strict condition (Ψ'). Yu. Egorov has proved the following important result (see [4], also [10]):

Theorem VI.2.- In order that the classical pseudodifferential operator B in X be subelliptic, it is necessary and sufficient that every point of $T^*X \setminus 0$ have an open neighborhood in which (Ψ'_k) is satisfied by B for some integer $k \geq 0$, depending on the point in question.

If this integer k can be chosen independently of the point, then, whatever x_o in X, Estimate (VI.13) will hold for $\delta = (k+1)^{-1}$ (and for a sufficiently small open neighborhood Ω of x_o).

The examples of boundary value problems for elliptic equations where boundary operators $\mathcal{B}$ of the kind considered in the present chapter occur are well-known: the standard ones are the so-called "oblique derivative" problems (see [2], [7]).

BIBLIOGRAPHICAL REFERENCES

[1] Agmon, S., Douglis, A. & Nirenberg, L.- Estimates near the boundary for solutions of elliptic partial differential equations satisfying general boundary conditions I, Comm. Pure Appl. Math., 12 (1959), 623-727; II, Comm. Pure Appl. Math., 17 (1964), 35-92.

[2] Boutet de Monvel, L.- Hypoelliptic operators with double characteristics and related pseudodifferential operators, Comm. Pure Appl. Math. 27 (1974), 585-639.

[3] Calderon, A. P.- Boundary value problems for elliptic equations, Proceed. Joint Soviet-American Symp. on Part. Diff. Equat., Novosibirsk (1963), 1-4.

[4] Egorov, Yu.- On subelliptic operators I, Uspekhi Mat. Nauk 30, 2 (1975) 57-114, II, Uspekhi Mat. Nauk 30, 3 (1975) 57-104.

[5] Egorov, Yu. & Kondratev, V. A.- The oblique derivative problem, Math. USSR Sb. 7 (1969), 139-169.

[6] Hörmander, L.- Pseudo-differential operators and hypoelliptic equations, Proceed. Symp. Pure Math. 10 (1967) 138-183.

[7] Hörmander, L.- A class of hypoelliptic pseudodifferential operators with double characteristics, Math. Ann. 217 (1975) 165-188.

[8] Folland, G. B. & Kohn, J. J.- The Neumann Problem for the Cauchy-Riemann Complex, Annals of Math. Studies Princeton 1972

[9] Lions, J. L. & Magenes, E.- Non-homogeneous Boundary Value Problems and Applications, Springer New York 1972.

[10] Menikoff, A.- Parametrices for subelliptic operators, Comm. Part. Diff. Equat. 2 (1977), 69-108.

[11] Nirenberg, L. & Treves, F.- On local solvability of linear partial differential equations, Part I: Necessary conditions, Part II: Sufficient conditions, Comm. Pure Appl. Math. 23 (1970) 1-38 & 459-510.

[12] Treves, F.- Topological Vector Spaces, Distributions & Kernels, Academic Press New York 1967.

[13] Treves, F.- Basic Linear Partial Differential Equations, Academic Press New York 1975.

[14] Treves, F.- Hypoelliptic partial differential equations of principal type. Sufficient conditions & Necessary conditions, Comm. Pure Appl. Math. 24 (1971), 631-670.

[15] Treves, F.- Winding numbers and the solvability condition (Ψ), Journal Diff. Geom. 10 (1974), 135-149.

Bressanone (Bolzano, Italy)

June 1977

GPSR Compliance

The European Union's (EU) General Product Safety Regulation (GPSR) is a set of rules that requires consumer products to be safe and our obligations to ensure this.

If you have any concerns about our products, you can contact us on ProductSafety@springernature.com

In case Publisher is established outside the EU, the EU authorized representative is:

Springer Nature Customer Service Center GmbH
Europaplatz 3
69115 Heidelberg, Germany

Batch number: 09490862

Printed by Printforce, the Netherlands